DATA IN THREE DIMENSIONS: A GUIDE TO ARCGIS® 3D ANALYST™

Heather Kennedy

D1501131

THOMSON

DELMAR LEARNING™

Australia   Canada   Mexico   Singapore   Spain   United Kingdom   United States

Data in Three Dimensions:
A Guide to ArcGIS 3D Analyst
Heather Kennedy

Vice President, Technology and Trades SBU:
Alar Elken

Editorial Director:
Sandy Clark

Senior Acquisitions Editor:
James DeVoe

Senior Development Editor:
John Fisher

Marketing Director:
Cynthia Eichelman

Channel Manager:
Fair Huntoon

Marketing Coordinator:
Casey Bruno

Production Director:
Mary Ellen Black

Production Manager:
Andrew Crouth

Production Editor:
Thomas Stover

Technology Project Specialist:
Kevin Smith

Editorial Assistant:
Katherine Bevington

Freelance Editorial:
Carol Leyba, Daril Bentley

Cover Design:
Cammi Noah

NOTICE TO THE READER

Publisher does not warrant or guarantee any of the products described herein or perform any independent analysis in connection with any of the product information contained herein. Publisher does not assume, and expressly disclaims, any obligation to obtain and include information other than that provided to it by the manufacturer.

The reader is expressly warned to consider and adopt all safety precautions that might be indicated by the activities herein and to avoid all potential hazards. By following the instructions contained herein, the reader willingly assumes all risks in connection with such instructions.

The publisher makes no representation or warranties of any kind, including but not limited to, the warranties of fitness for particular purpose or merchantability, nor are any such representations implied with respect to the material set forth herein, and the publisher takes no responsibility with respect to such material. The publisher shall not be liable for any special, consequential, or exemplary damages resulting, in whole or part, from the reader's use of, or reliance upon, this material.

About the Author

Heather Kennedy is a freelance technical writer and a GIS planner for the Contra Costa County government in California. She has been active in GIS for nine years as a writer, editor, and technical analyst. She is also the author of The ESRI Press Dictionary of GIS Terminology.

Acknowledgments

Thanks to John Fisher, Gary Amdahl, Kent Anness of Kentucky WRIS, Andrea Miller of the Department of Planning and Building in San Luis Obispo County, Tom McMurdo, Michael Kennedy, Lynda Gregory, Andrea Vanek, Damian Spangrud, Daril Bentley, Carol Leyba, and especially Tim Ormsby, whose work was the original basis for this book.

CONTENTS

Contents

Contents

INTRODUCTION

Data in Three Dimensions is a self-study workbook that teaches you how to use ESRI's 3D Analyst software, the ArcGIS extension for creating and displaying 3D data. This book is designed for people who are already familiar with the concepts of GIS, but who are new to the 3D modeling environment.

All of the exercises can be done with the 8.1 version of ArcView, ArcEditor, or ArcInfo, except for the animation exercises, which require versions 8.2 or higher.

> **NOTE:** *You must already have ArcGIS 3D Analyst installed to use this tutorial, as the book does not come with any trial software. The data sets for the exercises, however, are provided on the companion CD-ROM.*

Although you can do the exercises in any order, I recommend that you work through chapters 1 and 2 first, since instructions in later chapters are somewhat abbreviated, based on tasks performed in the first two chapters. Chapter 1 contains instructions for copying the tutorial data to your hard drive, loading the 3D Analyst extension, and making shortcuts on your desktop to ArcScene, ArcMap, and ArcCatalog.

About the Software

3D Analyst is primarily designed to create elevation data and display it in three dimensions, but provides additional analysis functions such as viewshed, surface area, and volume calculation. The more a map looks like the real world, the easier it is to understand. Experienced map readers can pretty easily interpret a flat map of elevation contour lines. But show the contour lines in 3D, or convert them to a triangulated irregular network (TIN), and the terrain needs no explanation, as the following illustrations point out.

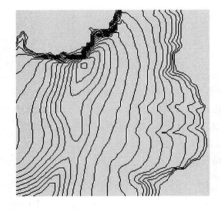

Contour lines in orthographic (2D) view.

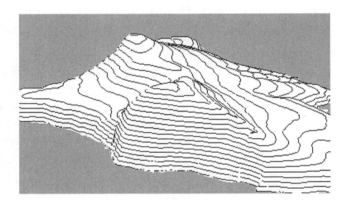

Contour lines in perspective (3D) view.

Contour lines converted to a TIN.

3D Analyst adds a new interface, ArcScene, to the ArcGIS application. ArcScene works like ArcMap, except that it lets you visualize and navigate around your data in three dimensions. ArcScene documents are saved as files with *.sxd* extensions, just as ArcMap documents are saved as *.mxd* files.

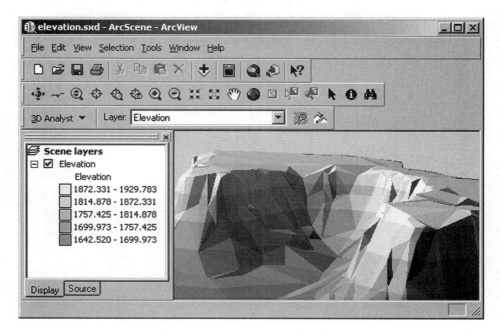

The ArcScene interface.

With ArcScene, you can create TIN and raster surface models from any vector elevation data, such as contour lines, GPS points, or survey points. You can drape images and vector features over surfaces, fly through your GIS data in 3D perspective, and make movies of your flights. You can convert 2D points, lines, and polygons to 3D and extrude them into lines, walls, and solids. You can calculate slope, aspect, hillshade, volume, and surface area; create contour lines; and determine visibility from any point on a surface.

An elevation raster in ArcMap.

The same elevation raster in ArcScene.

An aerial photograph draped over a digital elevation model (DEM).

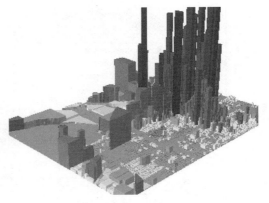

Parcels colored and extruded by land value.

TIN faces symbolized by slope.

A raster showing viewsheds from five observation points.

With the 3D Analyst extension loaded in ArcCatalog, you can preview TINs, rasters, 3D features, and ArcScene documents. You can also create metadata for your files, and make empty 3D feature classes for digitizing in ArcMap.

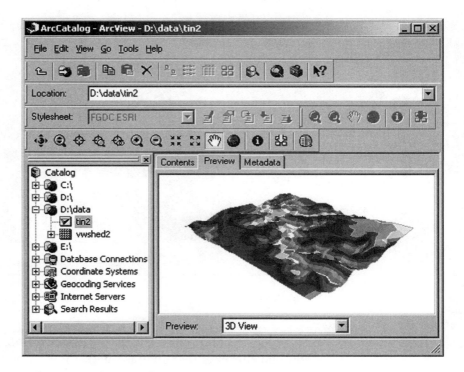

The ArcCatalog interface.

With the 3D Analyst extension loaded in ArcMap, you can do all of the same surface creation and analysis you can in ArcScene—the difference being that ArcScene can display it in 3D. 3D Analyst adds a few tools that only operate in ArcMap: you can determine lines of sight, create profiles graphs of lines on a surface, and digitize 3D features and graphics.

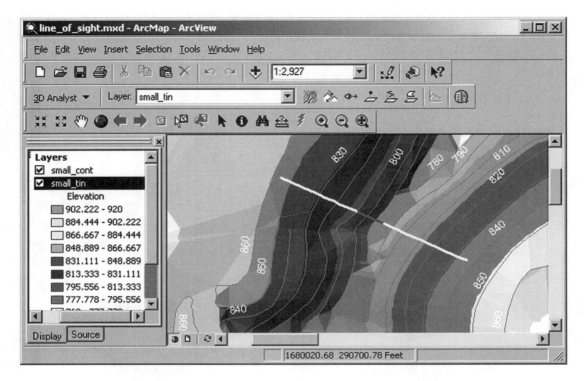

The ArcMap interface.

ArcScene, ArcMap, and ArcCatalog are customizable applications. You can make simple changes with the Customize dialog, or use the Visual Basic for Applications (VBA) editor that comes with ArcScene to create your own interfaces.

About the Companion CD-ROM

For information on the exercise data found on the companion CD-ROM, see the section "Load the Tutorial Data" in Chapter 1. It contains instructions for copying the tutorial data to your hard drive loading the 3D Analyst extension, and making shortcuts on your desktop to ArtScene, ArcMap, and ArcCatalog. See also the Note at the beginning of this introductions.

3D DATA OVERVIEW

Z VALUES

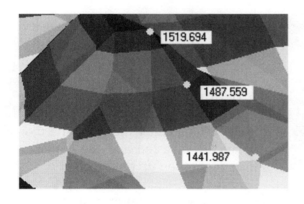

Three locations on the surface of a TIN, each labeled with its elevation (z) value in feet.

All geographical data contains horizontal *x,y* coordinate values. To work in three dimensions, you need data that contains *z* values as well. For each *x,y* location stored in a 3D data set, a *z* value is stored that represents an attribute other than that location's horizontal position. In a terrain model, the *z* value represents elevation, or height above sea level.

3D Analyst works primarily with raster, TIN (triangulated irregular network), and 3D vector feature data. Rasters and TINs are used to model surfaces, not just of terrain but of any phenomenon that varies continuously across an area, such as precipitation, chemical concentration, pollution dispersion, noise levels, population distribution, or soil pH.

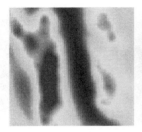  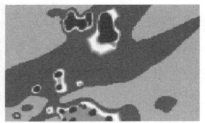

Temperature, elevation, and radioactivity are phenomena that can be modeled as continuous surfaces.

1 Agriculture
2 Brush
3 Grassland
4 Forest
5 Barren

Cells in a land-use grid. All cells with the same value are symbolized by the same color. While land use could also be represented by discrete vector polygons, vector data cannot represent values that change gradually, or continuously, over an area.

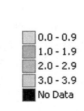

1.0	1.5	0.0
2.0	3.0	3.0
	0.5	1.0

0.0 - 0.9
1.0 - 1.9
2.0 - 2.9
3.0 - 3.9
No Data

Cell in a continuous grid, symbolized by value range.

RASTERS

A raster represents a surface as a rectangular grid of evenly spaced square cells. Each cell is the same size and has a unique row and column address. A cell can represent a square kilometer, a square meter, or a square centimeter. The smaller the cells, the more detailed the raster, and the larger the file space taken up by the grid.

Since the grid is uniform, its horizontal (x,y) coordinates don't need to be stored in each cell. Instead they are calculated from the x,y location of the lower-left cell in the grid. Each cell does, however, hold its own z value that represents a quantity or a category of phenomena such as elevation, crop yield, or reflected light.

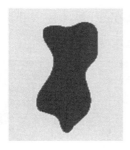

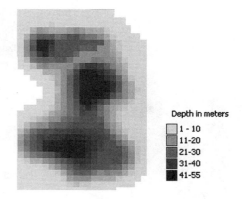

Depth in meters

☐ 1 - 10
▨ 11-20
▨ 21-30
▨ 31-40
■ 41-55

A vector polygon representing a lake. Its attributes are shape, area, circumference, and name. Since these attributes apply to the polygon as a whole, a measurement like depth can't be represented because it varies throughout the polygon.

A raster surface model of the lake. Each cell contains a z value representing the depth of the lake at that location.

Raster data is often divided into two categories: image and thematic. In an image, the surface phenomena is the reflection or emission of light, or some other band in the electromagnetic spectrum, and can be measured by camera or satellite.

An aerial photograph. Cells in this raster represent light reflected from the earth's surface.

When a phenomenon such as light is measured by a camera or a satellite, each cell's value correctly represents the light at that point on the surface. A thematic raster, however, represents a category or quantity of a phenomenon such as elevation, pollution, population, rainfall, or noise, and has to be measured by stationary instruments or by humans walking around with measuring equipment. Since no one can take a reading at every location, samples are taken instead, and a surface model is made. The model approximates the surface by interpolating the values between the sample points.

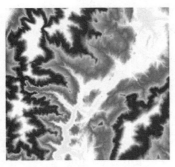

Elevation

High : 4361

Low : 438

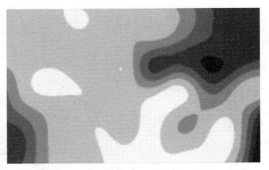

A thematic raster of elevation values. A few of the cells represent samples actually taken, but most of the values have been interpolated.

A raster surface of snow depth. Cell values have been interpolated from sample measurements.

3D Analyst uses the z value stored in each cell to display the raster in 3D. Elevation values are commonly shown, but any numeric cell value can be illustrated in three dimensions. Even though images and many thematic rasters don't contain elevation values, you can still display them in 3D by draping them over a 3D surface model with the same geographic extent.

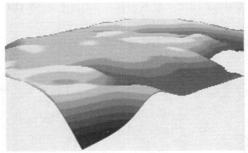

The snow depth raster draped over an elevation raster of the same area.

TINs

A triangulated irregular network (TIN) represents a surface as a set of irregularly located points, joined by lines to form a network of contiguous, non-overlapping triangles that vary in size and proportion. Each triangle node stores an x, y, and z value.

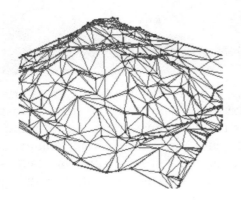

The structure of a TIN. Only the TIN edges and nodes are shown.

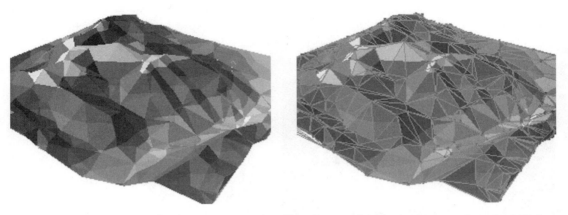

The structure of a TIN. Left: the TIN's triangles (faces) are colored to represent elevation. Right: Nodes, edges, and faces are shown together.

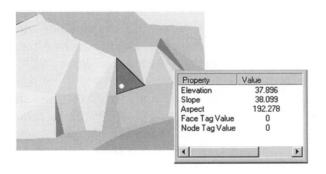

Property	Value
Elevation	37.896
Slope	38.099
Aspect	192.278
Face Tag Value	0
Node Tag Value	0

When you identify any point on the face of a TIN, the node x, y, z values are used to interpolate the elevation at that point. The node values are also used to calculate the slope and aspect of each triangle face.

Like rasters, the values in a TIN are interpolated from sample points. The sample points form the triangle nodes, and the interpolation (or triangulation, as it's generally called), consists of connecting the nodes by lines. Once the TIN is built, the elevation of any location on a TIN surface can be estimated using the x, y, and z values of the bounding triangle's vertices. The slope and aspect for each triangle face is also calculated.

Because the nodes can be placed irregularly over the surface, TINs can show greater detail where a surface is highly varied or where you want more accuracy. A TIN is only as good as the initial sample points taken; mountainous areas need many more samples per square unit than flat areas do in order to create an accurate terrain model.

TIN models are less widely available than raster surface models; they take longer to build and take up a lot more disk space. They are typically used for precise modeling of small areas.

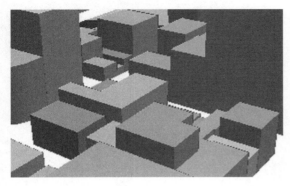

3D building polygons.

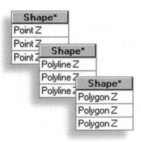

A 3D feature class shows a z value in its Shape field.

3D FEATURES

3D vector features, like their 2D counterparts, represent objects or clearly bordered areas such as buildings, land parcels, roads, power poles, and wells.

Like TINs, 3D features store *z* values along with *x,y* coordinates as part of their geometry. A point has one *z* value; lines and polygons have one *z* value for each vertex in the shape. You can identify 3D shapefiles and geodatabases by looking at the Shape field in their attribute tables.

Often, the *z* values in 3D features are used to represent an attribute other than height. For example, you might create a scene that shows city points extruded into 3D columns based on their population.

Population comparisons of cities in the U.S.

Now that you've had an introduction to 3D data structures, you're ready to learn the ins and outs of 3D Analyst. The next two exercises will teach you how to work with 3D data in ArcCatalog. Before you can do any exercises, however, you need to load the 3D Analyst tutorial data and add the ArcScene, ArcCatalog, and Arc-Map program icons to your desktop.

Load the Tutorial Data

Insert the 3D Analyst Tutorial Data CD into your CD drive. Copy the folder *GTK3D* to the drive of your choice. If you have the disk space, I recommend that you copy it directly under your *C:* drive, so that the full path name reads *C:\GTK3D.*

Once the *GTK3D* folder is copied to your hard drive, remove the CD and put it back in its sleeve. The *GTK3D* folder just contains data for the exercises—no programs—so you shouldn't need the CD again.

Add the Program Icons to Your Desktop

> **NOTE:** *ArcGIS 3D Analyst must be installed before you can create shortcuts to it on your desktop. If you have not installed the software, please see the ArcGIS 3D Analyst installation guide.*

1 On the taskbar of your desktop, click the Start menu. Move your cursor to Programs, then ArcGIS, and right-click Arc-Scene.

2 Choose *Send to* and then click Desktop (Create Shortcut). A shortcut to ArcScene is added to your desktop. (This procedure is for Windows XP; if you're running Windows 2000 or NT it may be a little different.)

3 If you don't already have icons on your desktop for ArcMap and ArcCatalog, you can use the same procedure to add them now.

EXERCISE 1: PREVIEW DATA IN ARCCATALOG

With 3D Analyst loaded in ArcCatalog, you can preview both 2D and 3D data in three dimensions. In this exercise you'll examine a TIN of Coletown, KY, and a 3D shapefile of contour lines.

Step 1
Start ArcCatalog

Double-click the ArcCatalog icon on your desktop. If you didn't make an ArcCatalog desktop icon, either see the instructions immediately above or click the Start menu, point to Programs, point to ArcGIS, and click ArcCatalog.

From the Tools menu, choose Options.

Click the General tab. At the bottom of the dialog, uncheck the box next to *Hide file extensions*.

Click OK.

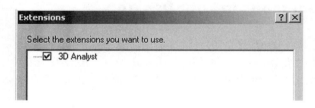

Step 2
Load the 3D Analyst Extension

From the Tools menu, choose Extensions. In the dialog, check the box next to 3D Analyst. Click Close.

Step 3
Load the 3D View Toolbar in ArcCatalog

From the View menu, click Toolbars and check the box next to 3D View Tools.

The 3D Analyst extension and the 3D View Toolbar are loaded. These tools let you view and navigate your data in 3D, query 3D features, and create perspective-view thumbnails.

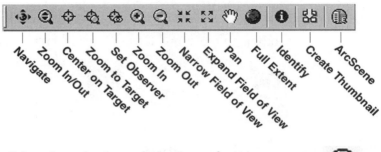

Other than the Launch ArcScene button

and the Create Thumbnail button,

all of these tools are also found on the 3D View toolbar in Arc-Scene.

Step 4
Look at ArcCatalog's Contents

In the Catalog tree on the left, navigate to the *GTK3D\Chapter01\ Data* folder. Click the plus sign next to the *Data* folder to open it.

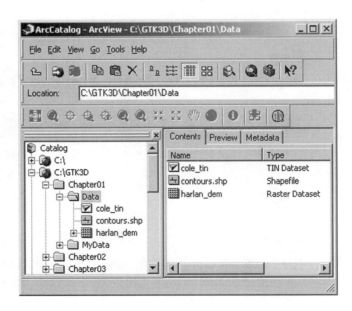

The *GTK3D\Chapter01\Data* folder contains one TIN, one raster, and one shapefile. As you can see, ArcCatalog uses the following symbols to represent TIN and raster data types:

TIN

Raster

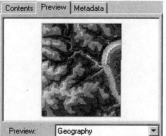

Step 5
Preview the TIN Data Set

Click *cole_tin* in the Catalog tree, then click the Preview tab above the display. The TIN is displayed in two dimensions (also called the orthographic or planimetric view).

Click the arrow next to the Preview menu below the display, and select 3D View.

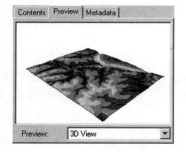

The TIN is displayed in three dimensions (perspective view). By default, the angle is from the southwest.

Step 6
Use the Navigate, Zoom, and Pan Tools

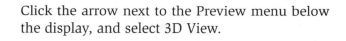

These are three tools you will use most to examine your 3D data.

Navigate

Zoom In/Out

Pan

The Navigate tool lets you rotate your data any direction around the center of the display. Click the Navigate tool.

Place your cursor over the center of the TIN in the display, hold down the left mouse button, and move the mouse in any direction. You can inspect the TIN from any angle, even from underneath.

To reset the view, click Full Extent.

Click the Zoom In/Out tool.

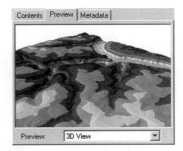

Place your cursor at the top center of the TIN in the display. Hold down the left mouse button, drag it down to the bottom of the scene, then release the mouse button. This zooms you in.

Now place your cursor at the bottom center of the display. Hold down the left mouse button, drag it up to the top of the scene, then release the mouse button. This zooms you out.

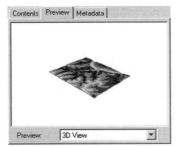

Click Full Extent again.

As in ArcMap, the Pan tool moves your data horizontally, vertically, or diagonally across the display. Click the Pan tool.

Place your cursor over the center of the display, hold down the left mouse button, and drag the cursor in any direction. When you're finished, click Full Extent.

Step 7
Pan and Zoom with the Navigate Tool

In addition to rotating your data in the display the Navigate button lets you pan and zoom without changing tools.

Click the Navigate tool again.

Place the cursor over the center of the display. Hold down both mouse buttons (with a three-button mouse, hold down the middle button) and drag the cursor in any direction. When you're comfortable panning with the Navigate tool, click Full Extent.

With the Navigate tool still selected, hold down the right mouse button. Drag it down in the display to zoom in, up to zoom out. When you're comfortable zooming with the Navigate tool, click Full Extent.

Step 8
Experiment with the Other Navigation Tools

The Narrow and Expand Field of View buttons and the Zoom In and Zoom Out buttons work the same way in ArcScene as they do in ArcMap.

Feel free to experiment with them. When you're finished, set the display back to Full Extent.

Step 9
Preview 3D Vector Features

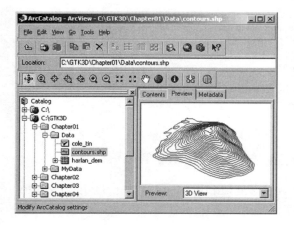

In the *GTK3D\Chapter01\Data* folder in the Catalog tree, click *contours.shp*. Make sure the Preview tab above the display is selected. Click the Navigate tool, and set the Preview menu below the display to 3D View.

(If you only see "contours," not "contours.shp" in the *Data* folder, click the Tools menu in ArcCatalog and choose Options. On the General tab, uncheck the box at the bottom that says "Hide File Extensions.")

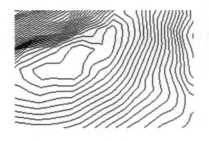

Put your cursor over the contour lines in the Preview display, hold down the left mouse button, and rotate them in different directions. Hold down the right mouse button to zoom in, then navigate some more.

Step 10
Compare the File Sizes of the Data Sets

Notice that the contour data rotates more quickly and smoothly than the TIN data. That's because the contour shapefile takes up much less disk space.

Click on the Metadata tab above the display. Enlarge the window so that you can see all three headings: Description, Spatial, and Attributes.

Click on Description. Scroll down to *Data storage and access information* and click on it. Scroll down some more and click on *Accessing the data*. Note that the file size of *contours.shp* is 0.184 MB.

Accessing the data
Size of the data: 0.184 MB
Data transfer size: 0.184 MB

In the Catalog tree to the left of the Preview display, click on *cole_tin*. Make sure the Metadata tab and the Description heading are selected. Scroll down again to *Data storage and access information*, click on it, and scroll further to *Accessing the data.*

Accessing the data
Size of the data: 2.687 MB
Data transfer size: 2.687 MB

The file size of *cole_tin* is 2.687 MB; over fourteen times larger than *contours.shp*. (Depending on a few factors having to do with metadata and thumbnail graphics, your file sizes may vary slightly.)

Step 11
Preview the 3D Shapefile Attribute Table

ArcCatalog uses the same icon for 2D and 3D shapefiles.

As you read earlier, one way to tell if a shapefile contains 3D features is to look at its attribute table. You can do this in ArcScene, ArcMap, or ArcCatalog. In the Catalog tree, click on *contours.shp* again. Click on the Preview tab above the display, and set the Preview menu below the display to Table.

FID	Shape*	CONTOUR
0	Polyline ZM	146
1	Polyline ZM	144
2	Polyline ZM	142
3	Polyline ZM	140
4	Polyline ZM	138
5	Polyline ZM	136
6	Polyline ZM	134
7	Polyline ZM	132
8	Polyline ZM	130
9	Polyline ZM	128

Contents | Preview | Metadata

Record: |◀ ◀ 1 ▶ ▶| Show: All Selected

Preview: Table

Look at the Shape field in the table. The expression *Polyline ZM* indicates that this is a 3D polyline feature. A 2D polyline feature would just have a Shape field value of *Polyline*.

Step 12
Query 3D Shapefile Attributes

You can also select individual 3D features and look at the attributes in ArcCatalog. In the Preview menu below the display, select 3D View. Notice that all of the 3D tools become active again.

Click the Navigate tool.

Put your cursor over the display, hold down the right mouse button, and drag it downward to zoom in on the contour lines.

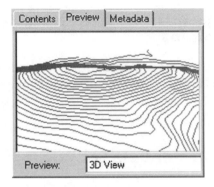

Click the Identify tool.

Click on one of the contour lines. ArcCatalog highlights it and the Identify Results box appears, listing the contour line's *x,y* location, its height (*z* value), and ID number.

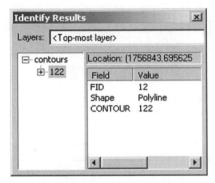

Click on a few more contour lines and note their elevations. When you're finished, close ArcCatalog.

In this exercise you loaded the 3D Analyst extension in ArcCatalog, previewed a TIN and a 3D shapefile, learned to use the 3D navigation tools, looked at 3D shapefile attributes, and examined metadata. In the next exercise, you'll preview a raster data set in ArcCatalog and create a layer from it.

EXERCISE 2: CREATE A LAYER FILE IN ARCCATALOG

In this exercise you'll preview a digital elevation model of Harlan, KY; create a layer file; symbolize it; and make a 3D thumbnail.

Step 1
Start ArcCatalog

Double-click the ArcCatalog icon on your desktop. If you didn't make an ArcCatalog icon, click the Start menu, choose Programs, then ArcGIS, then ArcCatalog.

Step 2
Preview a Raster Data Set

Navigate to the contents of your *GTK3D\Chapter01\Data* folder.

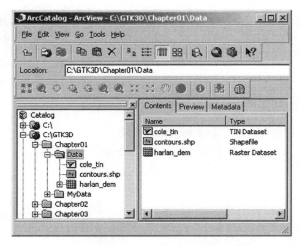

Click *harlan_dem* in the Catalog tree, then click the Preview tab above the display.

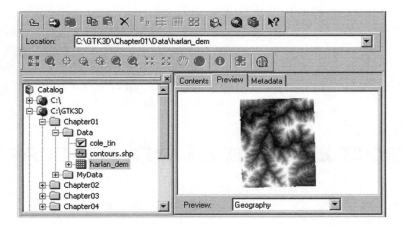

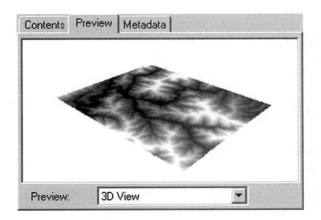

Change the Preview menu from Geography to 3D View. Notice that even though the DEM is shown in 3D perspective it still looks flat.

As you saw in Exercise 1, TINs and 3D vector features display their height values in ArcCatalog when you select 3D View. Rasters, however, are drawn as though they lie on a flat surface. In order to see the heights of an elevation raster in ArcCatalog you have to create a layer file *(.lyr)* from the raster and specify its 3D drawing properties.

Step 3
Create a Layer File from the Raster

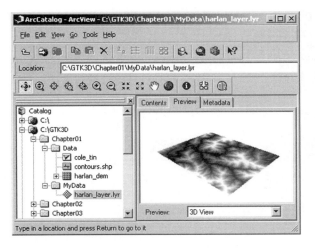

In the Catalog tree, right-click *harlan_dem* and choose Create Layer. Name it *harlan_laye*r and save it in your *GTK3D\Chapter01\MyData* folder (not the *Data* folder).

Again in the Catalog tree, open the *MyData* folder and click once on *harlan_layer.lyr* to highlight it.

Select the Preview tab above the display and choose 3D View from the Preview menu.

Harlan_layer.lyr is not a copy of *harlan_dem*; in fact, it's not a raster data set at all. It's a much smaller file that contains a copy of the display instructions for *harlan_dem*. You can't change the 3D viewing properties of *harlan_dem* in ArcCatalog, but you can change the 3D viewing properties of *harlan_layer*.

Step 4
Set Base Heights for harlan_layer

In the Catalog tree, right-click *harlan_layer* and click Properties.

In Layer Properties, click the Base Heights tab. Choose *Obtain heights for layer from surface.*

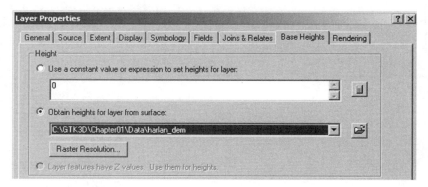

This setting uses the elevation values stored in *harlan_dem* to define the base heights of *harlan_layer*. You'll learn more about base heights in Chapter 2.

Click OK to close the Layer Properties dialog, and look at *harlan_layer* in 3D View.

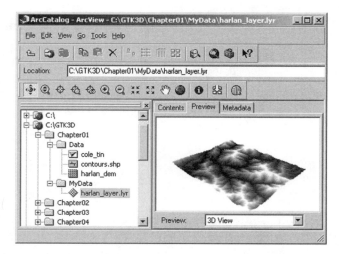

Harlan_layer draws in 3D, but you'll change the color scheme to better reveal its elevation levels.

Step 5
Change the Layer's Color Scheme

Right-click *harlan_layer* in the Catalog tree again and click Properties.

In Layer Properties click the Symbology tab. In the Color Ramp drop-down list, right-click on the color ramp itself (not on the drop-down arrow). Click Graphic View to uncheck it. This replaces the color ramp with its name.

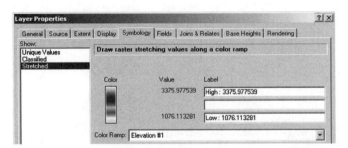

Click the Color Ramp drop-down arrow and scroll down until you see Elevation #1. Click to select it.

Click OK to close the Layer Properties dialog.

Preview *harlan_layer* again in 3D.

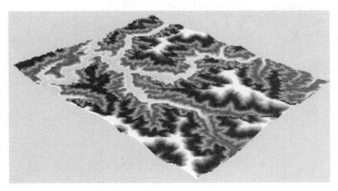

The elevation levels of this piece of Harlan County are much more apparent now, but you can improve the perspective further by adding shading to the surface.

Step 6
Add Shading to the Layer

Open Layer Properties again for *harlan_layer*.

Click the Rendering tab. In the Effects frame, check *Shade areal features relative to the scene's light position*.

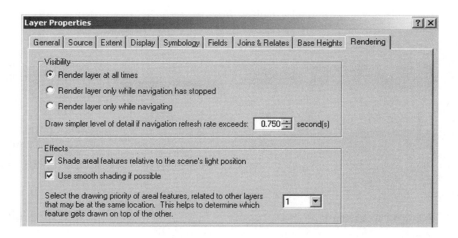

Click OK.

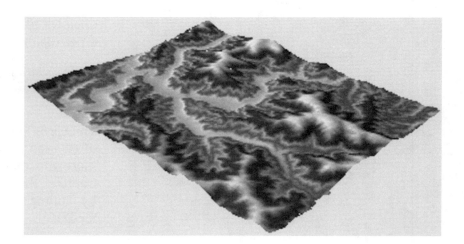

By illuminating the surface from a realistic sun angle, the elevation of the landscape stands out in greater relief. You'll learn more about illumination in Chapter 2.

Step 7
Create a 3D Thumbnail

In the Catalog tree, click once on *harlan_layer.lyr,* and select the Contents tab above the display.

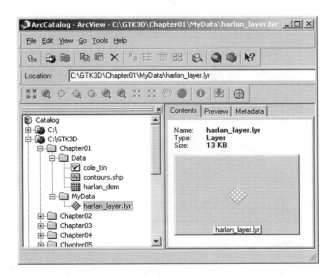

ArcCatalog shows the name, file type, and file size of *harlan_layer.* You'll create a 3D thumbnail to go along with the information.

Click on the Preview tab above the display and select 3D View from the Preview menu. The 3D tools are activated.

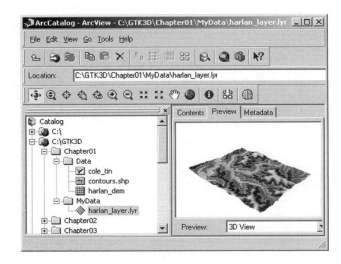

Click the Create Thumbnail button on the 3D toolbar.

Click the Contents tab again. The new *harlan_layer* 3D thumbnail is added to the file description.

Close ArcCatalog.

In this chapter you learned how to use the 3D Analyst navigation tools and how to preview, create, and symbolize 3D data in Arc-Catalog. Chapter 2 will introduce you to the ArcScene interface.

CHAPTER 2

3D DISPLAY IN ARCSCENE

Now that you know a bit about the types of data you can use in 3D Analyst, let's take a look at how ArcScene draws the data in 3D. ArcScene operates very much like ArcMap; you can add multiple data layers to a scene such as rasters, TINs, 2D, or 3D feature classes. ArcScene, however, lets you control how each layer is symbolized and positioned in 3D space. You can also change general settings that apply to the entire scene, such as sun illumination, background color, and vertical exaggeration.

SCENE PROPERTIES: BACKGROUND COLOR, ILLUMINATION, COORDINATE SYSTEM, AND VERTICAL EXAGGERATION

Background Color

By default, the background color in ArcScene is white. You can change it to whatever color you like, such as blue or black to simulate Earth sky or pinkish-orange to suggest the dust on Mars. You can choose colors from the Scene Properties palette or mix your own. You can also set a default background color for all new scenes.

Background color in ArcScene.

Illumination

Every 3D scene has a light source that determines which parts of
the surface are illuminated and which are in shadow. The position
of the light source is controlled by its altitude and azimuth settings.

The altitude is the angle, measured in degrees from 0 to 90,
between the light source and the horizon. An altitude of 0 degrees
puts the light source level with the horizon; an altitude of 90
degrees puts it directly overhead. ArcScene places the default altitude at 30 degrees.

The azimuth is the compass direction from which the sun or other
light source shines on the scene. It's measured clockwise in
degrees from 0 (due north) to 360 (also due north). The default
azimuth setting for the light source is 315 degrees, placing it in the
northwest.

You can change the azimuth and altitude of the light source and
alter the level of contrast. The illumination will affect all surfaces
and vector features in a scene.

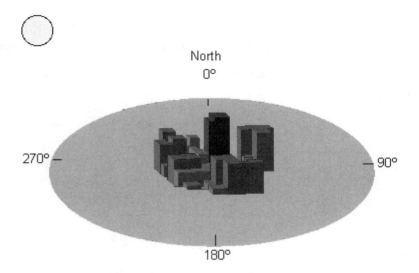

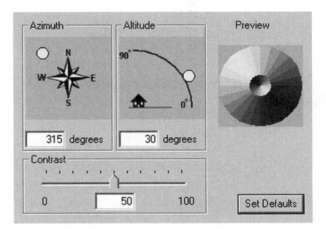

The default light source settings in ArcScene.

Coordinate System

When you open an empty scene, it has no coordinate system and adopts the coordinate system of the first data layer you add to it. This coordinate system overrides those of any subsequent layers added to the scene. If you add more layers that are in other systems, they are temporarily altered to match the one in the scene. This is called on-the-fly projection, and puts layers in different coordinate systems into spatial alignment.

You can select a different coordinate system for the scene at any time, and all layers in the scene will be temporarily reprojected. This will not alter the coordinate systems of the source data sets.

Like ArcMap, ArcScene expects coordinate system information to be stored with each data source. If it isn't, you'll need to know what system the data set uses so that you can give ArcScene the information. For a layer in a geodatabase, this information is part of the layer's metadata. For coverages, shapefiles, TINs, and rasters, it's stored in a separate file named after the data set, but with a *.prj* file extension (for example, *parcels.prj*). If the *.prj* file hasn't been provided with the data set but you have the coordinate system information, you can create one with ArcCatalog.

If no coordinate information is associated with a data set, Arc-Scene will test the coordinate values to see if they fall within the range of −180 to 180 degrees for *x* values and −90 to 90 degrees for *y* values. If they do, ArcScene assumes that these are geographic coordinates of latitude and longitude. If the values aren't in this range, ArcScene considers them planar *x,y* coordinates.

Vertical Exaggeration

You can exaggerate the vertical appearance of surfaces by multiplying the *z* units in a scene by a number. A vertical exaggeration of 2 multiplies all *z* values by 2, an exaggeration of 0.4 multiplies all *z* values by 0.4, and so on. It's often used to emphasize slight changes in elevation on a surface that looks flat because of its large extent. Conversely, a fractional vertical exaggeration can smooth choppy surfaces.

Vertical exaggeration is also used to bring *z* values into proportion with *x,y* values when their units are different. For example, the *x,y* values of a raster may be in UTM (Universal Transverse Mercator) meters but the elevation measurements may have been taken in feet.

Like a scene's illumination and coordinate system, vertical exaggeration affects all the layers in a scene. It is just a visual effect, and doesn't influence measurement or analysis.

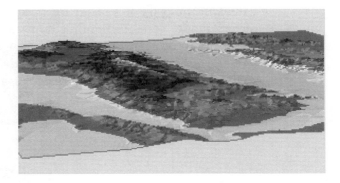

A TIN with no vertical exaggeration.

The same TIN with a vertical exaggeration of 3 applied to the scene.

3D LAYER DISPLAY PROPERTIES: BASE HEIGHTS, EXTRUSION, AND SHADING

In addition to the global settings that affect all layers in a scene, each layer has its own set of 3D drawing properties.

Base Heights

The base height of a layer is the elevation at which it is drawn in a scene. As you saw when you loaded 3D Analyst in ArcCatalog, TINs and 3D features are automatically drawn in 3D, but rasters and 2D features look flat. The same holds for ArcScene. ArcScene gets the base heights for TINs and 3D features from the z values in their nodes and vertices, respectively. In order to draw rasters in 3D, however, you have to tell ArcScene where to get the z values to use for base heights. ArcScene doesn't automatically use the values in the cells of a raster for base heights because quite often rasters model something other than elevation, for example in an aerial photo or a grid of pollution dispersion. (Non-elevation rasters and 2D vector features are instead "draped" over another 3D surface, such as an elevation raster or a TIN, whose *z* values are used for base heights.)

An elevation raster and a 2D feature layer of rivers. When they're first added to ArcScene their base heights are set to 0, so they lie flat.

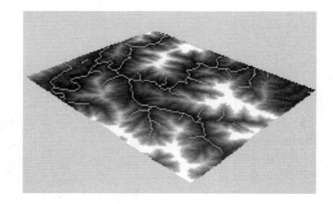

The elevation raster's base heights have been set to its cell values, so ArcScene displays it in 3D. Base heights have not yet been set for the features, so they lie underneath the raster.

The base heights for the 2D river features have been set to the base heights of the raster. Now the rivers drape over the surface.

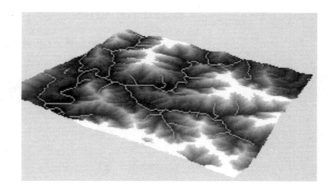

Extrusion

Extrusion extends 2D vector features vertically from their base height to a point you specify. You can extrude points into vertical lines, lines into walls, and polygons into solids.

2D polygons become blocks when extruded into three dimensions.

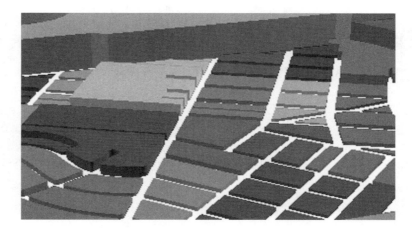

Shading

You can also control whether or not individual layers are shaded by the illumination settings for the scene. Shading gives layers a more realistic appearance. Although the Illumination setting is naturally the same for all layers in a scene (the light source comes from one altitude and one angle), whether a layer "participates" in the shading is up to you.

An unshaded elevation raster.

The same raster, shaded according to the sun's altitude and azimuth settings in the scene.

In this chapter you'll work with ArcScene. You'll learn to set background color, illumination, vertical exaggeration, and coordinate system; you'll set base heights for an elevation raster, an aerial photo, and 2D vector features; and you'll extrude 3D polygon features into solid shapes.

EXERCISE 1: SET BACKGROUND COLOR AND ILLUMINATION IN ARCSCENE

When you open a new document in ArcScene, 3D Analyst places the sun in the northwest at 30 degrees above the horizon, so that north- and west-facing surface slopes are lit, and south- and east-facing slopes are in shadow. In this exercise you'll learn how to change these settings.

Step 1
Start ArcScene

Double-click the ArcScene icon on your desktop. If you didn't make an ArcScene icon in Chapter 1, click the Start menu, point to Programs, point to ArcGIS, and click ArcScene.

Step 2
Load the 3D Analyst Toolbar

In ArcScene, the 3D Analyst extension should be loaded by default. Check this by clicking on the Tools menu, then Extensions. The box next to 3D Analyst should be checked.

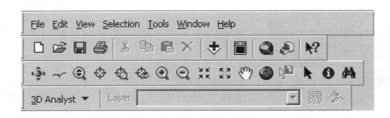

From the View menu, choose Toolbars, then check the box next to 3D Analyst. This loads the 3D Analyst menu and toolbar.

Step 3
Add hill_tin to the Scene

Click the Add Data button in ArcScene.

Navigate to your *GTK3D\Chapter02\Data* folder. Double-click *hill_tin.lyr* to add it to the scene.

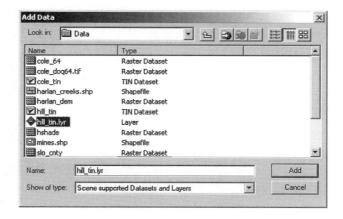

Hill_tin.lyr is a layer file that uses the TIN surface *hill_tin* for its source. *Hill_tin* is also in your *GTK3D\Chapter02\Data* folder.

Click the Navigate tool

and hold down the left mouse button while moving the cursor around in the scene. Hold down the right mouse button and move the cursor downward in the scene to zoom in to the hill.

Step 4
Set a Background Color for the Scene

In the Table of Contents, right-click Scene Layers and choose Scene Properties. Click the General tab.

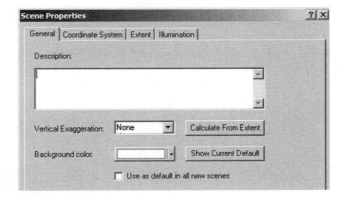

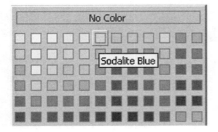

Click the Background Color drop-down arrow, and choose Sodalite Blue.

Click OK.

The scene's background color changes to a light blue.

Step 5
Add a Navigation Marker

Click the Full Extent button.

Click the Add Data button again.

Open your *GTK3D\Chapter02\Data* folder. Double-click *south_ end.lyr* to add it to the scene.

The yellow marker sits on the southern slope of the hill to help you keep your bearings as you navigate. Notice that your default vantage point at full extent is from the southwest.

Step 6
Change the Level of Contrast in the Scene

 Click the Navigate tool.

Hold down the left mouse button in the scene and move the surface from right to left until you are looking at the hill from the south instead of the southwest. Hold down the right mouse button to zoom in, and hold down both mouse buttons together to pan. The scene should look something like this:

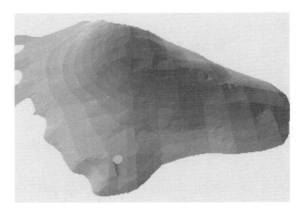

Move the ArcScene application window to the right side of your computer screen.

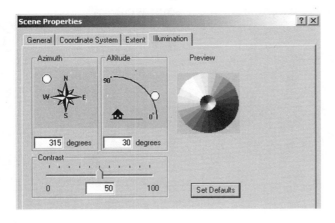

In the Table of Contents, right-click Scene Layers and choose Scene Properties. (You can also double-click the words *Scene Layers* to bring up the Scene Properties.) Move the Scene Properties dialog box to the left side of your screen so you can see the dialog box and the scene.

Click the Illumination tab.

Drag the Contrast slider bar back and forth between 0 and 100. The slider bar controls the amount of shading applied to the surface. As the contrast is increased, the shadows deepen, emphasizing variation on the surface.

The default Contrast setting is 50. In the Contrast box, highlight the value and type *75*. Leave the Scene Properties dialog open.

Step 7
Change the Sun's Altitude and Azimuth

You can change the sun's position by moving the sun icon with your cursor or by typing in values.

In the Azimuth frame, drag the sun icon from the northwest to the west, then south, then east. Alternating between lighting the northern and southern sides of the hill will give you a good idea of how azimuth is modeled in ArcScene.

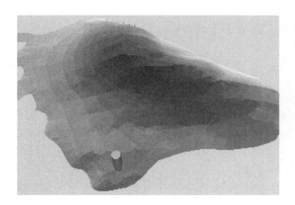 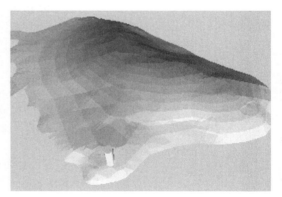

Put the sun in the northwest in the Azimuth frame.

In the Altitude frame, drag the sun icon up to 90 degrees and then down to zero. The higher the sun, the more directly each face of the TIN's surface is lit. At 90 degrees (midday) most of the detail of the hill is obliterated. At 0 degrees (sunset) it is almost completely in shadow.

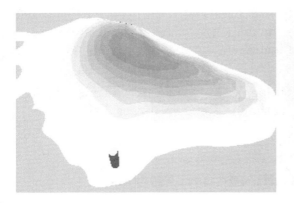

Step 8
Experiment with Illumination Settings

Click Set Defaults to restore the original settings. Feel free to play with the Azimuth, Altitude, and Contrast settings.

If you want to navigate around the scene, you have to click OK to close the Scene Properties dialog first. To reopen the dialog, right-click Scene Layers in the Table of Contents.

Step 9
Close ArcScene

Close the Scene Properties dialog. If you'd like to save the Arc-Scene document, go to the ArcScene File menu and click Save. Give it a name of your choice and save it in your *GTK3D\ Chapter02\MyData* folder (not the *Data* folder). The *.sxd* extension indicates that the file is an ArcScene document, just as an *.mxd* extension indicates an ArcMap document.

Close ArcScene.

EXERCISE 2: SET VERTICAL EXAGGERATION IN ARCSCENE

Vertical exaggeration is used to emphasize small changes in elevation, or to bring z values into proportion with x,y values when they're in different units. In this exercise you'll apply vertical exaggeration to terrain in San Luis Obispo County, California.

Step 1
Start ArcScene and Add a Raster Layer

Double-click the ArcScene icon on your desktop, or open Arc-Scene from the Start menu.

Click the Add Data button in ArcScene.

Navigate to your *GTK3D\Chapter02\Data* folder. Double-click *slo_cnty.lyr* to add it to the scene.

Step 2
Look at the Layer File and Its Data Source

Notice that the layer file's name in ArcScene's Table of Contents is *slo_cnty*, not *slo_cnty.lyr*, even though in the Add Data dialog box it is called *slo_cnty.lyr*. That's because ArcScene names the layer file after its data source when you add it to the scene. In this case, the data source is a digital elevation model (DEM) called *slo_cnty*. (If you installed the *GTK3D* folder somewhere other than your *C:* drive, you may see a small red exclamation mark next to the layer in the Table of Contents. In that case, read on and your question will be answered.)

Click the Add Data button again, and select the Details button on the Add Data toolbar.

This view of the data tells you that *slo_cnty* is a raster data set, and that *slo_cnty.lyr* is a layer file.

slo_cnty   Raster Dataset
slo_cnty.lyr   Layer

Now select the Thumbnails button.

Scroll down in the Add Data window until you can see *slo_cnty* and *slo_cnty.lyr*.

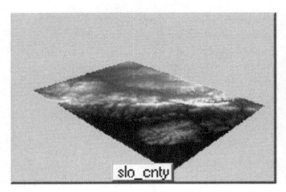   

Slo_cnty is the raw raster data set. *Slo_cnty.lyr* is a file containing a set of instructions telling ArcScene, ArcMap, or ArcCatalog to draw *slo_cnty* with a certain color scheme, shading, and base height setting. When you add *slo_cnty.lyr* to ArcScene, it references *slo_cnty*. For this reason, if you were to move *slo_cnty* to another folder *slo_cnty.lyr* would be unable to reference it. (If this happens, ArcScene gives you the option to "Set Data Source" by right-clicking the layer file in the Table of Contents. For more information on repairing data sources, search the ArcGIS Desktop Help for "Repairing the data source in a layer.")

Close the Add Data dialog without adding any data.

Step 3
Navigate the Scene

Click the Navigate tool.

Use the left and right mouse buttons to navigate and zoom to the surface of *slo_cnty.lyr*. Use both buttons together to pan.

This area of San Luis Obispo is about 1,200 square miles—an area so large that the 963-foot difference between the highest and lowest elevation points looks small.

Step 4
Set a Vertical Exaggeration for the Scene

Click the Full Extent button.

Move ArcScene to the right side of your desktop to make room for the Scene Properties dialog box.

Open the Scene Properties dialog. You can do this in any of three ways: by choosing Scene Properties from the View menu, by right-clicking *Scene layers* in the Table of Contents and choosing Scene Properties, or by double-clicking the words *Scene layers* in the Table of Contents.

Move the Scene Properties dialog to the left of your screen so that you can see the layer in ArcScene.

Click the General tab.

Highlight "None" in the Vertical Exaggeration box and choose 5.

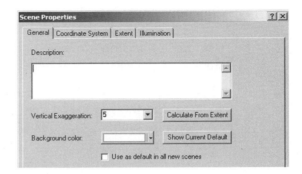

Click OK.

Use the Navigate tool to examine the surface. The hills are a little taller, the valleys a little deeper.

Step 5
Have ArcScene Calculate Vertical Exaggeration

Open the Scene Properties dialog again. This time click Calculate From Extent.

ArcScene calculates a vertical exaggeration for you based on the extent of all data displayed in the scene. In this case, it is about 27.5.

Click OK.

San Luis Obispo is beginning to look like the Alps. Notice that in the Table of Contents the elevation values have not changed.

As mentioned earlier, vertical exaggeration is a visual effect applied to all layers in a scene. It does not change data values or influence analysis.

Step 6
Change Vertical Exaggeration and Color Scheme

Open the Scene Properties dialog again. This time, highlight the value and type *75*.

Click OK.

In ArcScene, click the Full Extent button to center the layer in the display.

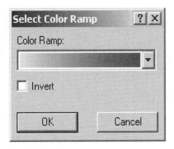

In the Table of Contents, click once on the color ramp for *slo_cnty*. This brings up the Select Color Ramp dialog.

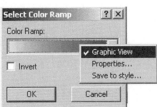

Right-click the color ramp in the dialog box, and uncheck Graphic View. This gives you the names of the color ramps.

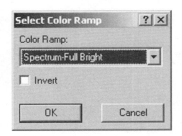

Scroll down through the ramp descriptions and choose Spectrum-Full Bright.

Click OK in the Select Color Ramp dialog.

Step 7 Navigate Again

Use the Navigate tool to take a look at your crazy new landscape. If you always wanted a set of Magic Rocks as a kid, now you can instantly grow your own in ArcScene.

Step 8
Close ArcScene

When you're finished navigating, close ArcScene. Click No when you're asked if you want to save changes.

EXERCISE 3: APPLY A COORDINATE SYSTEM TO A SCENE

When you add a layer to an empty scene, the scene adopts that layer's coordinate system. If you add subsequent layers that use different coordinate systems, ArcScene temporarily projects the layers to fit the scene. This does not affect the coordinate systems of those layers' source data. In fact, you can change the coordinate system of a scene as many times as you like and you won't affect the source data of the layers in the scene.

In this exercise you will add a TIN of an area near Coletown, KY, to a scene. You'll change the scene's coordinate system from feet to meters, and change the vertical exaggeration to compensate for visual impression created by displaying z values in meters instead of feet.

Step 1
Start ArcScene

Double-click the ArcScene icon on your desktop, or open Arc-Scene from the Start menu.

Click the Add Data button in ArcScene.

Navigate to your *GTK3D\Chapter02\Data* folder. Double-click *cole_tin* to add it to the scene.

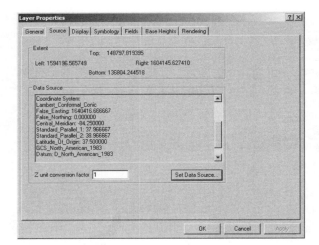

Step 2
Look at cole_tin's Coordinate System in Layer Properties

You can find a layer's coordinate system information in several places. Right-click *cole_tin* in the Table of Contents and select Properties. In the Layer Properties dialog, click the Source tab.

Cole_tin uses a Lambert Conformal Conic projection based on the NAD 1983 datum. There is more to know about *cole_tin*'s coordinate system, though.

Click OK to dismiss the Layer Properties dialog.

Step 3
Look at cole_tin's Coordinate System in ArcCatalog

In ArcScene, click the ArcCatalog button.

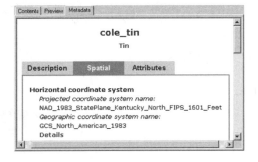

In ArcCatalog, navigate to your *GTK3D\ Chapter02\Data* folder. Select *cole_tin*.

Click the Metadata tab. Within the Metadata window, click the Spatial tab.

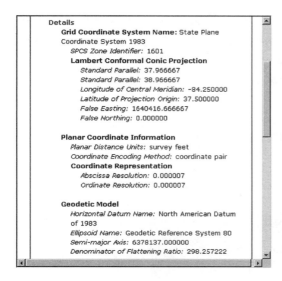

Cole_tin is in the NAD 1983 State Plane coordinate system.

Click the Details button to see all of the coordinate system metadata.

The State Plane coordinate system is designed for large-scale (small-area) U.S. mapping that divides the 50 states into over 120 numbered zones, each with its own projection parameters. The Lambert Conformal Conic projection is used for states that are longer in the east-west direction (such as Kentucky and Tennessee), and the Transverse Mercator projection is used for states that are longer in the north-south direction (such as Vermont and Illinois). Notice that the units are survey feet, which is something the Source tab in the Layer Properties dialog doesn't tell you.

Close ArcCatalog.

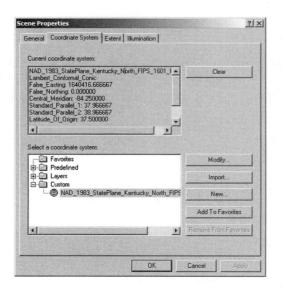

Step 4
Look at cole_tin's Coordinate System in Scene Properties

In the Table of Contents, right-click Scene Layers and choose Scene Properties.

Click the Coordinate System tab.

Because *cole_tin* is the first layer added to the scene, you know the scene is using *cole_tin*'s coordinate system. Like the Metadata tab in ArcCatalog, the Scene Properties dialog gives you a little more information than Layer Properties does.

When you change the scene's coordinate system in a few minutes you'll see that the scene no longer reflects *cole_tin*'s coordinate system.

Close the Scene Properties dialog.

Step 5
Zoom in to the Kentucky River

Use the Navigate tool to zoom in, pan, and get a good look at the banks of this particular bend in the Kentucky River.

Click the Identify tool.

Click a few places along the river. Notice that it is about 540 feet above sea level. The *x* and *y* coordinates are also in feet. Take a look at the range of numbers you get with the Identify tool at various locations.

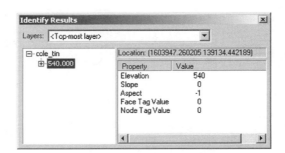

When you're finished examining the surface with the Identify tool, close the Identify Results box.

Step 6
Change the Scene's Projection

Now you are going to project the scene into the UTM zone that is appropriate for the location of *cole_tin*.

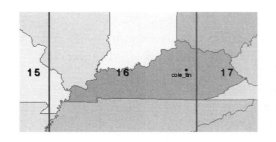

The UTM system divides the globe into 60 zones running north-south, each spanning six degrees of longitude. Each zone has its own central meridian and is split by the equator into north and south sections. As you can see from the map at left, *cole_tin* lies in zone 16.

Open the Scene Properties dialog (right-click or double-click Scene Layers in the Table of Contents, or choose Scene Properties from the View menu).

The Coordinate System tab should still be selected. In the window under *Select a coordinate system*, click Predefined, then click Projected Coordinate Systems, then UTM, then NAD 1983.

You're choosing UTM NAD 1983 because the Kentucky State Plane coordinate system *cole_tin* uses is also based on the NAD 83 datum.

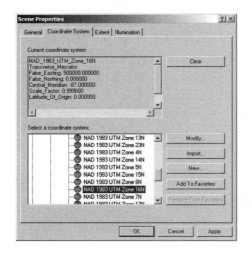

Scroll down in the list of NAD 1983 UTM zones and select Zone 16N. (The N is for North, indicating that Kentucky is north of the equator.)

Click OK to change the scene's coordinate system from Kentucky State Plane to UTM Zone 16N.

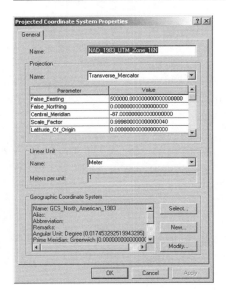

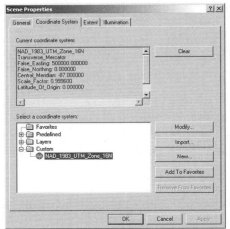

Step 7
Examine the Scene with a UTM Projection

Use the Navigate tool to look at *cole_tin* again. It looks a lot lumpier. Zoom to the bend in the river. The banks have become a canyon. What's going on?

The explanation can be found by looking at the properties of the new UTM projection.

Double-click Scene Layers to bring up Scene Properties again. The Coordinate System tab should be selected.

NAD_1983_UTM_Zone_16N is the scene's current projection. You may recall that when the scene was in Kentucky State Plane the units—feet—were part of the name of the coordinate system and so showed up in the title under "Current Coordinate System." You can't tell from the title here what units the coordinate system is in, though, so you'll have to dig a little deeper.

With *NAD_1983_UTM_Zone_16N* selected, click Modify.

Under Linear Unit notice that the units for this projection are meters, not feet.

Click Cancel to close the Projected Coordinate System Properties dialog without making any modifications. (If you accidentally click OK, don't worry about it—you'll just have a duplicate listing of the current coordinate system.)

Click OK to close the Scene Properties dialog.

Step 8
Identify Locations in the River Again

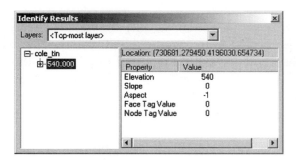

When you changed the scene's coordinate system from State Plane to UTM, the *x,y* units were successfully changed from feet to meters, so horizontal measurements taken in either projection are properly converted. Changing the coordinate system doesn't change the *z* units, however.

Click the Identify tool. Click some locations in the riverbed again.

As we saw earlier, the elevation of the Kentucky River at this point is about 540 feet. If the z units had been converted to meters, they would report the height of the river at about 180, or about a third of 540. But the Identify results still report a value of 540. We know that they're feet, but ArcScene thinks they're meters, so the z units of *cole_tin* look three times taller than they should.

Step 9
Compensate for the Feet-to-Meters Discrepancy by Changing the Scene's Vertical Exaggeration

There are several ways to fix the way *cole_tin* appears in the scene. One way is to go back to the Modify button in the Projected Coordinate System Properties dialog and change Meter to Foot. Another way is to change the vertical exaggeration of the scene.

Double-click Scene Layers in the Table of Contents to bring up the Scene Properties dialog.

Click the General tab.

Highlight "None" in the Vertical Exaggeration box, and type in the value *0.3048*. This effectively converts the z units in the scene from feet to meters.

Click OK in the Scene Properties dialog.

Voila! *Cole_tin* looks normal again.

Vertical exaggeration is just a visual effect, remember. If you use the Identify tool, the *z* measurements will be reported in feet, and the *x,y* measurements will be reported in meters. That's because ArcScene gets the *x,y* units from the scene's projection, but gets the *z* units from the coordinate system of *cole_tin*'s source data, which as you know is State Plane Feet. To see the elevation reported in meters, you'd need to reproject the actual *cole_tin* data set to a new coordinate system.

Step 10
Close ArcScene

Close the Identify Results box and close ArcScene. Say No when asked if you want to save your changes.

EXERCISE 4: SET 3D LAYER PROPERTIES FOR AN ELEVATION RASTER

As you learned in the beginning of this chapter, you have to set the base heights of a raster layer when you add it to a scene. In this exercise you'll add a raster to ArcScene and set its base heights, shading, and symbology.

Step 1
Open ArcScene

Double-click the ArcScene icon on the desktop, or open ArcScene from the Start menu.

Click the Add Data button in ArcScene.

Navigate to your *GTK3D\Chapter02\Data* folder. Double-click *cole_64* to add it to the scene.

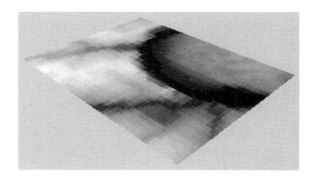

Step 2
Examine the Data

In the Table of Contents, right-click *cole_64* and choose Open Attribute Table.

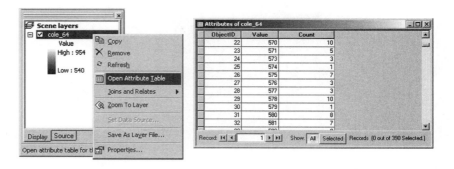

Cole_64 is a small portion of a DEM of Coletown, KY. You worked with a TIN made from that DEM in the last exercise. It has 390 records, one for each unique elevation value in the raster. The elevations are in State Plane Feet and are stored in the Value field. For each record, the Count field shows how many cells have that elevation.

Close the table.

Step 3
Set Base Heights

When you add a raster to ArcScene, you will generally have to set its base heights, shade it, and symbolize it. You do all of these through the Layer Properties dialog.

Move ArcScene to the right side of your screen so that you can still see *cole_64* when you open the Layer Properties dialog box.

Right-click *cole_64* in the Table of Contents and choose Properties. Move the dialog to the left of your screen.

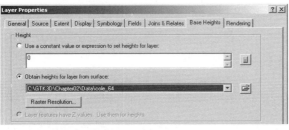

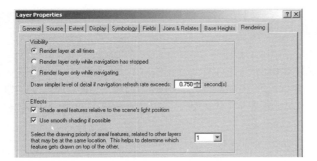

Click the Base Heights tab.

As you may remember, base heights for rasters are set to 0 by default, so the layer starts out flat.

Choose *Obtain heights for layer from surface* and pick *cole_64* from the drop-down list (it should be the only data set listed).

Click Apply, and leave the dialog box open.

Now ArcScene displays the layer according to the elevation values you just examined in the Value field of *cole_64*'s attribute table.

Step 4
Set the Shading for cole_64

In the Layer Properties dialog, click the Rendering tab.

In the Effects panel, check the box next to *Shade areal features relative to the scene's light position. Use smooth shading if possible* is checked automatically.

Click Apply again.

Now the raster is shaded from the northwest. It looks a bit dark, so in a minute you'll raise the sun to a higher altitude.

Step 5
Change the Symbology of cole_64

Click the Symbology tab.

Right-click the color ramp so that you can uncheck the Graphic View box.

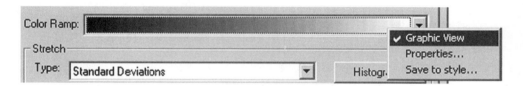

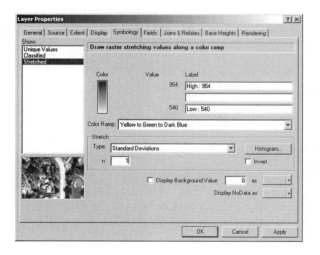

Scroll through the list of color ramps and choose Yellow to Green to Dark Blue.

Under Stretch, change the n value to 1. Increasing the value of n shades more cells with colors from the middle of the ramp, whereas decreasing the value of n shades more cells with colors at each end of the ramp.

Click OK. This closes the Layer Properties dialog box. Now *cole_64* is displayed in 3D, shaded, and symbolized according to its elevation values.

Step 6
Add a Background Color and Change the Sun's Altitude

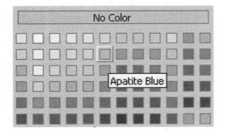

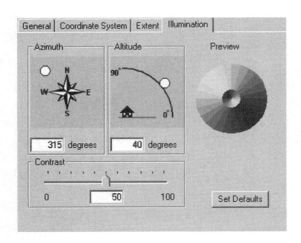

In the Table of Contents open the Scene Properties dialog by double-clicking Scene layers or by right-clicking *Scene layers* and choosing Scene Properties.

Click the General tab.

Click the arrow next to the Background color box and choose Apatite Blue.

Click Apply, and leave the Scene Properties dialog open.

Now click the Illumination tab.

In the Altitude box, highlight the value and type in *40*.

Click OK to close the Scene Properties dialog.

Step 7
Navigate the Data Set

Cole_64 is more brightly illuminated for viewing. Click the Navigate tool and zoom in and around the raster.

Step 8
Save the ArcScene Document

From the File menu, choose Save. Navigate to your *GTK3D\ Chapter02\MyData* folder. Give the ArcScene document a name of your choice and click Save. The ArcScene document is saved in your *Chapter02\MyData* folder.

Close ArcScene.

EXERCISE 5: SET 3D LAYER PROPERTIES FOR A RASTER IMAGE

In the last exercise you added an elevation raster to ArcScene and set its base heights so that it would display in 3D according to its own elevation values. In this exercise you'll drape an image raster over the elevation raster.

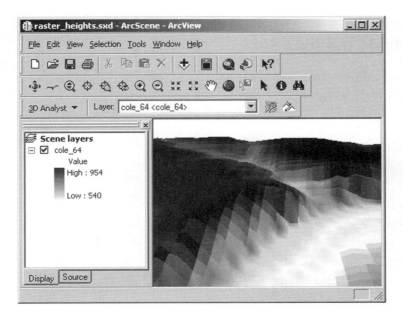

Step 1
Open an ArcScene Document

Start ArcScene. From the File menu choose Open, and navigate to your *GTK3D\Chapter02\Data* folder.

Double-click *raster_heights. sxd*. This is a copy of the ArcScene document you made in the last exercise.

Step 2
Add an Aerial Photograph

Click the Add Data button.

You should still be in your *GTK3D\Chapter02\Data* folder.

Double-click *cole_doq64.tif* to add it to the scene.

Click the Full Extent button.

The new raster's base heights are set to 0 by default when it is first added to ArcScene, so it looks flat.

Cole_doq64 is a digital orthophoto quad, an aerial photograph that has been geometrically rectified to account for distortions owing to the camera's tilt and uneven terrain. This means that distances and directions between landmarks in the photo are proportional to distances and directions measured on the ground.

This doesn't mean that *cole_doq64.tif* has elevation values in its cells, however. It's a black-and-white image, so its cells contain grayscale values ranging from 0 (black) to 255 (white). In Arc-Scene, you can borrow the base heights from an elevation raster or a TIN and use them as the base heights for an image, for 2D vector data, or for other rasters that represent phenomena besides elevation. The elevation surface is typically present in the scene, but it can just be a layer you browse to on disk.

Step 3
Set Base Heights for the Aerial Photo

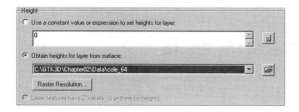

In the Table of Contents, right-click *cole_doq64.tif* and choose Properties.

Click the Base Heights tab. Click *Obtain heights for layer from surface* and in the drop-down list choose *cole_64* (not *cole_doq64. tif*).

Click OK to dismiss the Layer Properties dialog box.

The aerial photo is draped over the elevation raster. It looks strange because the base heights of the photo are set to the base heights of the elevation raster, so they're competing for the same display space.

There are a couple of ways to deal with this. The simplest is to turn off the elevation layer.

In the Table of Contents, uncheck the box next to *cole_64*.

Now you can see the aerial photo by itself.

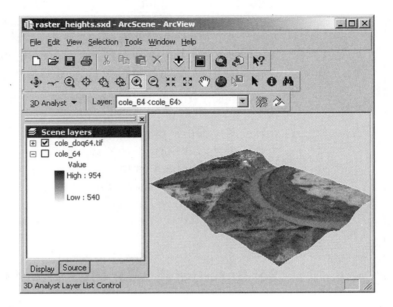

Step 4
Add a Height Offset to the Photo

Suppose you want to see the photo on top, but you'd still like to be able to look at the elevation raster. You can see both by adding an offset to the image layer.

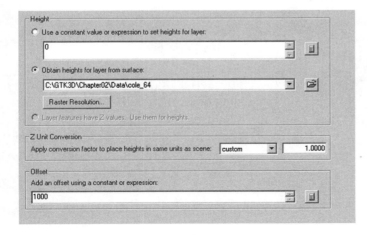

Turn *cole_64* back on in the Table of Contents.

Double-click *cole_doq64.tif* to bring up its layer properties. The Base Heights tab should be selected.

In the Offset panel, type *1000*. This raises the image layer 1,000 units—in this case, feet—above the elevation raster.

Click OK.

Now you can easily examine both surfaces. Use the Navigate tool to pan and zoom around the layers.

Step 5
Save the ArcScene Document

From the File menu, choose Save As. Give the document a new name and save it in your *GTK3D\Chapter02\MyData* folder. Then close ArcScene.

EXERCISE 6: SET BASE HEIGHTS FOR A 2D VECTOR LAYER

In the last exercise you used an elevation raster to set base heights for an image. In this exercise you'll use an elevation raster to set the base heights for 2D vector features.

Step 1
Start ArcScene and Add an Elevation Raster

Double-click the ArcScene icon on your desktop, or open Arc-Scene from the Start menu.

Click the Add Data button.

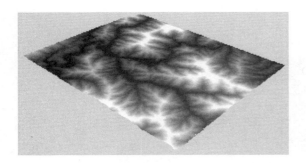

Navigate to your *GTK3D\ Chapter02\ Data* folder.

Double-click *harlan_dem* to add it to the scene.

You worked with this DEM of Harlan County, KY, in Chapter 1.

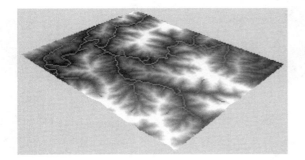

Step 2
Add a 2D Shapefile

Click the Add Data button again. Double-click *harlan_creeks.shp* to add it to the scene.

The raster's base heights are set to 0, and the creeks are 2D polyline features, so both layers lie flat.

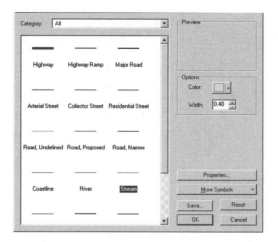

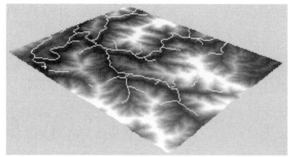

Step 3
Change the Line Symbol for harlan_creeks

In the Table of Contents, click on the line symbol underneath *harlan_creeks*.

This brings up the Symbol Selector dialog. Click on the line symbol above *Stream* to select it.

Click OK. Now the creeks show up more clearly against the DEM.

Step 4
Examine the Attribute Table of the harlan_creeks Shapefile

In the Table of Contents, right-click *harlan_creeks* and choose Open Attribute Table.

The shapefile has 69 records. You can tell that it's a 2D shapefile because the records in the Shape field say *Polyline*. If it were a 3D shapefile, the records would say *PolylineZM*.

Close the attribute table.

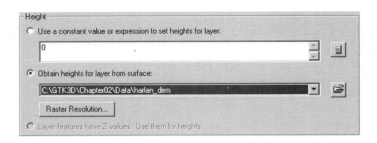

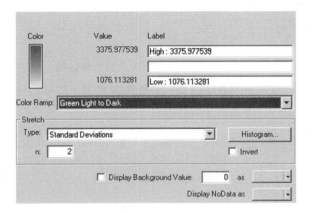

Step 5
Set the Elevation Layer's Properties

In the Table of Contents, right-click *harlan_dem* and choose Properties.

Click the Base Heights tab.

Check *Obtain heights for layer from surface* and choose *harlan_dem* as the source in the drop-down list.

Click the Rendering tab.

In the Effects box, choose *Shade areal features relative to the scene's light position*.

Finally, click the Symbology tab, and right-click the color ramp so that you can uncheck Graphic View.

Scroll through the color ramp descriptions and choose Green Light to Dark.

Click OK to activate the settings and close the Layer Properties dialog.

Harlan_dem is displayed in 3D, but the creeks can't be seen.

Step 6
Set Base Heights for the Creeks Layer

Use the Navigate tool to zoom in on the scene, and take a look underneath *harlan_dem*.

The creeks haven't had their base heights set yet, so they're hidden under *harlan_dem*.

In the Table of Contents, right-click *harlan_creeks* and choose Properties.

Click the Base Heights tab.

Click *Obtain heights for layer from surface* and select *harlan_dem* for the source of heights in the drop-down menu (it should be your only choice).

Click OK.

Click the Full Extent button.

Now the 2D creeks are using *harlan_dem*'s elevation values as base heights, so you can see them on top. Each vertex in the polyline shapefile adopts the elevation value from the corresponding cell in *harlan_dem*.

Step 7
Navigate the Scene

Zoom in and look at the places where the creeks flow through the hills. In certain areas and from certain angles the creeks disappear into the terrain, giving it a stitched effect.

If you turn off *harlan_dem* you can see that the creeks are continuous, but the lines between the vertices are being buried by the variations in the raster layer.

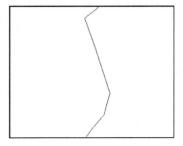

Step 8
Change harlan_dem's Drawing Priority

In the last exercise you saw how to offset layers that are competing for the same 3D space. You can also set a layer's drawing priority so that it displays above or below other layers.

Move ArcScene to the right side of your screen.

In the Table of Contents, right-click *harlan_dem* and choose Properties.

Click the Rendering tab, and move the Layer Properties dialog to the left, out of the way of the scene.

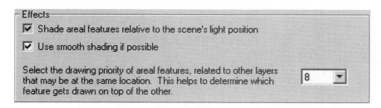

In the Effects frame of the dialog, change the layer's drawing priority to 8. A value of 1 is the highest priority, and is the default for all layers. 10 is the lowest drawing priority setting. By setting *harlan_ dem* to 8, you tell it to draw underneath *harlan_ creeks.shp*.

Click OK.

The elevation raster now has the lowest drawing priority in the scene. More of the creeks appear.

Step 9
Offset the Shapefile

In the Table of Contents, right-click *harlan_creeks* and choose Properties.

Click the Base Heights tab, and move the Layer Properties dialog to the left of the scene.

In the Offset panel, type *25* into the box. This raises the creek layer 25 feet above the raster layer.

Click OK.

Now the stitching is gone, and the creeks display fully on top of the raster.

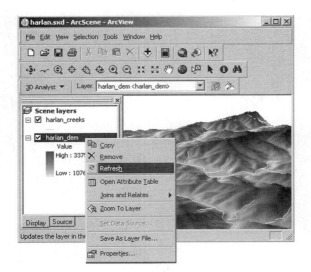

Step 10
Navigate Again

Click the Full Extent button.

Use the Navigate button to look at the creeks from various angles. You should be able to see them on top of the raster layer.

In the Table of Contents, right-click *harlan_dem* and choose Refresh.

(If the raster appears "holey" you can set its drawing priority to a higher number, and give the creeks more of an offset to compensate. As a hint, the first operation is performed in the Layer Properties dialog for *harlan_dem*. The second is performed in the Layer Properties dialog for *harlan_creeks*.)

Step 11
Save the ArcScene Document

From the File menu, choose Save As.

Give the document a name of your choice and save it in your *GTK3D\Chapter02\MyData* folder.

Close ArcScene.

EXERCISE 7: EXTRUDE 2D VECTOR FEATURES

In the last exercise you used an elevation raster to set the base heights for 2D line features. In this exercise you'll use an elevation raster to set the base heights for 2D polygons, and you'll extrude them by depth values provided in their attribute table. You may remember from the discussion in Chapter 2 that when vector features are extruded points become vertical lines, lines become walls, and polygons become solids.

Step 1
Start ArcScene and Add an Elevation Raster

Double-click the ArcScene icon on your desktop, or open it from the Start menu.

Click the Add Data button.

Navigate to your *GTK3D\Chapter02\Data* folder. Double-click *slo_cnty* to add it to the scene.

This is the elevation raster you used earlier, covering an area of about 1,200 square miles in San Luis Obispo County, CA.

Step 2
Set Base Heights for the Raster Layer

Right-click *slo_cnty* in the Table of Contents and choose Properties.

Click the Base Heights tab. Click *Obtain heights for layer from surface*, and select *slo_cnty* as the source for heights.

Click OK.

Step 3
Set Vertical Exaggeration and Background Color for the Scene

In the Table of Contents, double-click Scene Layers to bring up the Scene Properties.

Click the General tab. In the Vertical Exaggeration box, choose 10.

In the Background color box, choose Apatite Blue.

Click OK.

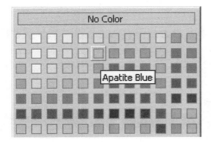

Step 4
Navigate the Layer

Click the Navigate tool.

Right-click the mouse and pull it down in the scene to zoom in. Take a look around the layer.

Step 5
Add the 2D Polygons

Click the Add Data button again. You should still be in the *GTK3D\Chapter02\Data* folder. Double-click *mines.shp* to add it to the scene.

Click the Full Extent button.

You can't see the mines because their base heights haven't been set, and they're hidden under *slo_cnty*.

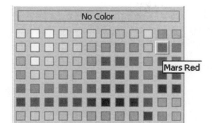

Step 6
Change the Color of the Mine Polygons

In the Table of Contents, turn off *slo_cnty*.

Right-click the polygon symbol under *mines* to bring up the color palette.

Choose Mars Red.

Step 7
Examine the Mines

Now you should be able to see the mines. Use the Zoom In tool to draw a rectangle around a few that are clustered.

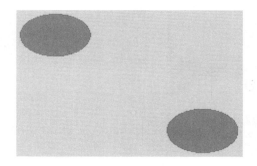

Use the Navigate tool to examine them more closely. Remember that by holding both mouse buttons down together you can pan the scene.

These polygons are circles that were created by buffering a point shapefile of mine locations in San Luis Obispo. Each has a diameter of about 2,350 feet, or almost half a mile. They don't reflect the actual acreage of the mines any more than the points did, but as polygons they are easier to see from a distance.

Step 8
Look at the Mine Attributes

In the Table of Contents, right-click *mines* and choose Open Attribute Table.

Attributes of mines

FID	Shape*	MINE1	MINETYPE	N_ELEV
12	Polygon	WHALE ROCK PIT	OPEN PIT	300
6	Polygon	TIBER CANYON SAND PIT	OPEN PIT	450
7	Polygon	TEMPLETON/ORMONDE	GRAVEL	740
14	Polygon	SYCAMORE ROAD PIT	GRAVEL	850
5	Polygon	SANTA RITA STONE QUARRY	QUARRY	140
16	Polygon	SANTA MARGARITA QUARRY	QUARRY	100
20	Polygon	ROCKY CANYON QUARRY	QUARRY	107
13	Polygon	PATCHETT PIT	OPEN PIT	540

Record: 0 — Show: All | Selected | Records (0 out of 23 Selected.)

Mines.shp has 23 records. If you scroll to the right of the attribute table, you'll see a field called *N_ELEV* that contains the negative elevation, or lowest depth, of each mine.

Also look at the field called *MINETYPE*. There are three types of mines: quarry, open pit, and gravel.

Close the attribute table.

Step 9
Set the Base Heights for the Mines

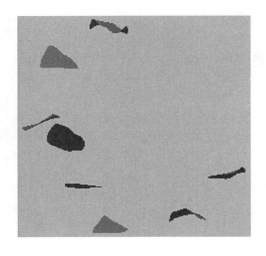

Slo_cnty should still be turned off in the scene, and you should still be zoomed in closely to a few of the mine polygons.

In the Table of Contents, right-click *mines* and choose Properties.

Click the Base Heights tab, click *Obtain heights for layer from surface*, and select *slo_cnty* as the source for heights.

Click OK.

The polygons are using the base heights from *slo_cnty*, so they're draped over the surface of the raster. Use the Navigate tool to look at how the shapes of the circles are changed by the terrain. From underneath they look like gently drifting red potato chips.

Step 10
Extrude the Polygons

Click the Full Extent button, and turn *slo_cnty* back on in the Table of Contents.

Right-click *mines* and choose Properties.

Click the Extrusion tab.

When you looked at the attribute table for *mines.shp* you saw the field called *N_ELEV* that contains the mine depths. You'll extrude the mines by their depth values.

Click the box next to *Extrude features in layer.*

Click the Calculator button to open the Expression Builder dialog.

In the Fields box, click *N_ELEV* to add it to the Expression box. Because you want to extrude the mines downward, you need to put a minus sign (a hyphen) in front of the expression. The extrusion value will be applied to each feature's minimum height.

Click OK in the Expression Builder.

Click the Rendering tab.

Check the box to use smooth shading. When polygon or line features are extruded, this box is unchecked by default. (When point features are extruded, no shading is available.)

Click OK to close the Layer Properties dialog.

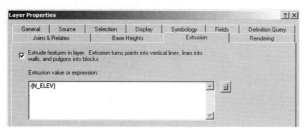

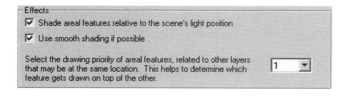

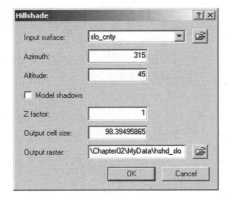

Now the base heights are set for both layers, and the mines are extruded. Before you take a look down under, let's give the raster some hillshading for better contrast against the mines.

Step 11
Make a Hillshade of slo_cnty

This step will give you a preview of some of the surface analysis you'll be performing in later chapters.

Click the 3D Analyst drop-down menu in ArcScene. Select Surface Analysis, and then click Hillshade.

In the Hillshade dialog, the input surface is *slo_cnty*.

Click the Browse button next to the Output raster box and navigate to your *GTK3D\Chapter02\ MyData* folder.

Name the new layer *hshd_slo*.

Click OK, and wait for the hillshade to be processed. When it's finished it will be added to the scene.

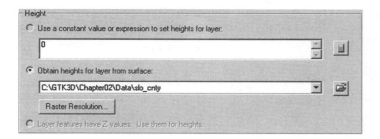

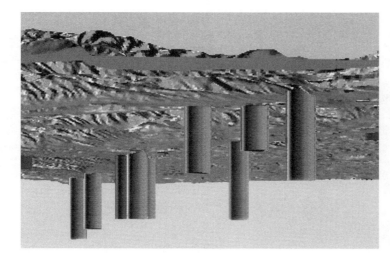

Step 12
Set Base Heights for the Hillshade of slo_cnty

In the Table of Contents, right-click *hshd_slo* and choose Properties. Click the Base Heights tab and check *Obtain heights for layer from surface.*

From the drop-down list, choose *slo_cnty* (not *hshd_ slo*) as the source for base heights.

Click OK to close the Layer Properties dialog.

In the Table of Contents, turn off *slo_cnty* so it won't compete for display space with *hshd_slo*.

Use the Navigate tool to zoom in on the scene, and take a look underneath the hillshade layer. The mines are extruded downward by the values in the *N_ELEV* field in their attribute table. (The units are feet.)

The vertical exaggeration of the scene affects the mine depths as well as the raster heights. Without such a large vertical exaggeration, the mines would barely poke beneath the surface.

Step 13
Symbolize the Mines by Type

Let's give the mines a bit more personality by coloring them according to their type.

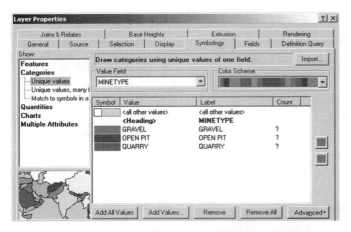

In the Table of Contents, right-click mines, choose Properties, and click the Symbology tab.

In the Show box, choose *Categories: Unique values.*

In the Value Field drop-down list, choose *MINETYPE.*

Click the Add All Values button. Uncheck the box next to the topmost symbol that says *All other values.*

Choose a style you like in the Color Scheme box.

Click OK to close the Layer Properties dialog.

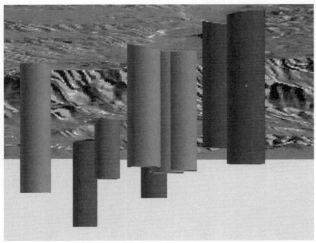

Step 14
Close ArcScene

When you're finished checking out the mines, close ArcScene. If you'd like to save the document, give it a name of your choice and save it in your *GTK3D\Chapter02\MyData* folder.

In this chapter you learned how to set 3D display properties for scenes and layers. In Chapter 3 you'll learn how to animate your data, fly through terrain, and make movies.

CHAPTER 3

3D NAVIGATION AND ANIMATION

TARGETS AND OBSERVERS

You already know how to use the Navigate tool to pan, zoom, and move around your data in a 3D scene. In this chapter you'll learn to specifically control your perspective in a scene, fly through terrain, create an animation sequence, and save it as a movie file.

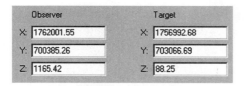

The star indicates the location of the target in the scene. The x, y, *and* z *values of both target and observer are shown in the box above.*

When you change your perspective with the Navigate tool, you are telling ArcScene to change the position of the target (what you're looking at) and the observer (you). Officially, the observer is the point of the scene viewer relative to the data in the scene, and the target is the center point around which a scene viewer moves when you navigate. The relative position between the target and observer defines your 3D perspective.

Target and observer points each have an x, y, and z value. You can set specific target and observer locations, either by clicking in the scene with the Target and Observer tools or by typing x, y, and z coordinate values in the scene's View Settings.

75

ANIMATED ROTATION

You can also make your data spin around its axis at various speeds, while you pan, zoom in and out, or navigate. Rotating a scene is a good "hands-off" way to get an overview of the data.

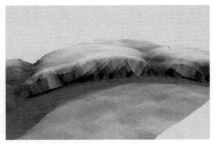

Animated rotation sets the data spinning around its axis.

SIMULATING FLIGHT

The Fly tool lets you navigate through data from a bird's-eye view. When the Fly tool is activated, the target and observer coordinates change continually, giving the illusion of flight over a motionless landscape.

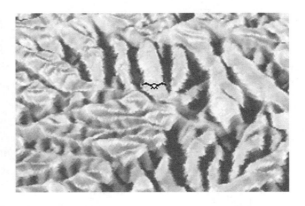

Flight simulation through the terrain of a hillshade map.

CREATING ANIMATION

After you've had some experience flying through your data and placing targets, you'll create an animation file by linking scenes. ArcScene lets you make movies that show changes in scene properties (illumination, background) perspective (navigation, rotation, flight) or any change you apply to your data. These films can be saved as ArcScene animation documents or converted to *.avi* files.

EXERCISE 1: SET TARGETS AND OBSERVERS

When you're looking at data in ArcScene, you might want to know what you can see from a specific position, or what an object looks like from different angles. In this exercise you'll learn about targets and observers, orthographic and perspective view, roll angle, and pitch.

Step 1
Start ArcScene and Add Data

Open ArcScene from your desktop, and click the Add Data button.

Navigate to your *GTK3D\Chapter03\Data* folder. Add *3d_bldgs.lyr* and *tar_tin.lyr* to the scene.

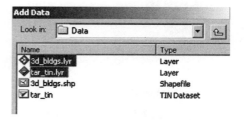

You've added a TIN layer and a layer of buildings to the scene. One of the features is a bright yellow 20-foot pole representing an observation tower, from which you'll be surveying the scene.

Step 2
Add a Background Color

In the Table of Contents, right-click Scene Layers and choose Scene Properties.

In Scene Properties, click the General tab. Click the drop-down box next to *Background color*, and choose Sodalite Blue.

Click OK.

Step 3
Examine the Scene

Use the Zoom In and Navigate tools to take a look at your surroundings.

Step 4
Center on a Target

Click the Full Extent button, and then zoom in to the cluster of blue and red buildings.

Click the Center on Target tool.

This tool centers the view on any point you click in the scene. Click on the roof of the red building as indicated in the graphic at left.

Notice that the scene pans to make the roof-top the center (or near center) of the scene.

Step 5
Navigate Around the Target

Click the Navigate tool.

Hold down the left mouse button, and move the cursor around in the scene. Notice that the surface rotates around the point you set as the target.

Now hold down the right mouse button, and drag the cursor up and down in the scene. Notice that the target remains the focus while zooming. (The only way to change the target is to hold down both mouse buttons, which will pan to a different location.)

You can also center on a target while you're using the Navigate tool. To do so, hold down the Ctrl key, and then left-click anywhere in the scene. The scene centers on the new location.

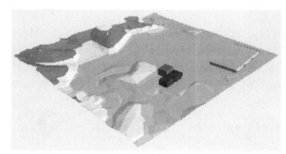

Step 6
Zoom to a Target

Click the Full Extent button. The target is reset to the center of the surface.

Click the Zoom to Target tool.

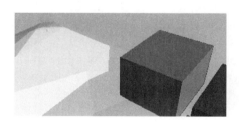

This tool lets you set and zoom in to a target with one click. Click the rooftop of the red building again, as best you can with the scene set to its full extent. ArcScene zooms in to the top of the building.

If you keep using the Zoom to Target tool while you're this close, you'll lose sight of your target. The Zoom to Target tool works best from a greater distance.

Click the Navigate tool, and click the Full Extent button again. This time, zoom to a target while you're using the Navigate tool. To do so, hold down the Ctrl key, and then right-click on the desired spot. The scene will center and zoom in on the new location.

Step 7
Set an Observer

Click the Full Extent button. Use the regular Zoom In tool to make your scene looks like the graphic at left.

You're going to target the rooftop again, but you'll also set a point of observation.

Click the Center on Target tool.

Click the red building. The scene centers on the rooftop. Make sure you can still see the yellow pole on the facing slope of the hill. (If you can't, use the Full Extent, Zoom In, and Center on Target tools again until you can see the red building and the yellow pole.)

Now click the Set Observer tool.

Click on the yellow pole on the hillside. Arc-Scene zooms in and relocates your point of observation. You should be looking at the buildings, now, from approximately the perspective shown at left.

If you don't get a point of view close to the one above, run through Step 7 again. It takes a little practice and a little luck.

Step 8
Set Target and Observer Coordinates

Clicking target and observer points in the scene is quick and convenient, but if you know their coordinates you can create a more exact model of the visual landscape. The View Settings dialog box allows you to control x, y, and z positions of both observer and target precisely, and reports the line-of-sight distance between them.

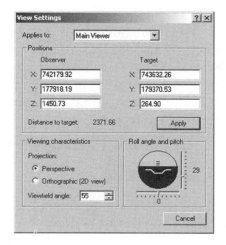

Click the Navigate tool, and click the Full Extent button to return to the original scene.

Open View Settings from the View menu.

In it you see the coordinates of the current observer (you) and target (the middle of the scene). You also see that it's in perspective or 3D view, and that the roll angle and pitch are, respectively, set to 0 and 29.

Step 9
Change Target and Observer Coordinates

Move the View Settings dialog away from the scene so that you can see the data.

Type the following coordinates into the Observer and Target fields, so that the dialog looks like the box at left.

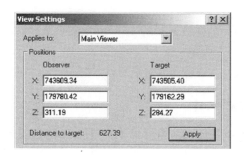

When you're finished typing in both sets of numbers, click Apply.

3D Analyst calculates the values, zooms to the observer position and faces the target location.

The target and observer coordinates you typed in are the same locations you've been working with. The resulting scene should look pretty familiar:

Notice that the distance from your perspective—the top of the yellow pole—to the rooftop of the closest red building is 627.39 meters. This is a 3D measure, which means that it accounts for vertical as well as horizontal distance. There is a direct line of sight between target and observer. If the pole were a cell tower and you were to climb to the roof of the building to talk on your cell phone, you would probably get excellent reception.

Step 10
Create a Bookmark

In the next few steps you'll be moving the scene around, and you may lose the current perspective. Since you probably don't want to type in the target and observer values all over again, now's a good time to set a bookmark.

Under the View menu, choose Bookmarks, and click Create.

Since you're only making one bookmark, you can leave the title "Bookmark 1." To get back to this view at any time, choose Bookmarks from the View menu and select it.

Click OK to close the 3D Bookmark dialog.

Step 11
Adjust Angle and Projection

The View Settings dialog should still be up. If not, open it again from the View menu. Drag it so that it's not in the way of the scene.

Look at the fields in the Viewing Characteristics panel. The projection is Perspective (3D), and the Viewfield angle is around 55 (yours may be different).

In ArcScene, experiment with the Narrow Field of View and Expand Field of View buttons.

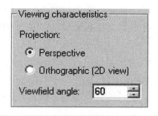

The Viewfield angle in the View Settings dialog changes each time to reflect the new distance. Narrow the field of view to 25.

You can also type values directly into the Viewfield angle box, or use the up and down arrows. Type in the value *60*. The scene pulls back to the larger viewing angle.

Step 12
Change Projection

Under Viewing Characteristics, click Orthographic.

ArcScene shows the full extent of the scene in two dimensions, just like ArcMap. Orthographic scenes are limited to 2D navigation, so many of your 3D tools no longer work. The Narrow/Expand Field of View buttons, the Set Observer tool, the Navigation tool, and the Fly tool are disabled. You can still use the Center on and Zoom to Target tools, the Pan tool, and the Zoom In/Out tools to get a closer look. When you're done, click the Perspective button.

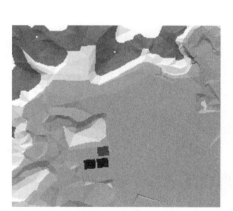

Step 13
Change Roll Angle and Pitch

You can also adjust the pitch to elevate your point of observation. If observer and target are at the same height, the pitch is zero. If the observer is higher, the pitch is positive; if the observer is lower, the pitch is negative.

If you changed your perspective several times in Step 12 and lost the scene from Bookmark 1, select it now from the View menu.

In the View Settings dialog, drag the vertical Pitch slider up and down to change your elevation with respect to the target. The pitch ranges from +89 to –89 degrees. At +89, your observation point is almost directly above the building. At –89, it's underneath the surface of the TIN.

Slide the vertical bar back to 1 or 2 degrees.

Now slide the horizontal bar to the left and right. This controls the roll angle, and is similar to looking out of the cockpit of a plane while it tips its wings left and right. This angle ranges from 180 to –180 degrees.

Step 14
Close ArcScene

Feel free to experiment with the View Settings dialog from various perspectives. When you're finished, close ArcScene. If you'd like to save the scene, give it a name of your choice and save it in your *GTK3D\Chapter03\MyData* folder.

EXERCISE 2: ANIMATED ROTATION AND THE VIEWER MANAGER

Animated rotation puts your data in motion so that you can look at it without using the mouse or keyboard. In this exercise you'll animate a TIN surface from San Luis Obispo, CA. You'll also use the Viewer Manager to look at the surface from three simultaneous perspectives.

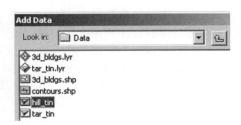

Step 1
Start ArcScene and Add Data

Open ArcScene, and click the Add Data button. Navigate to your *GTK3D\Chapter03\Data* folder. Add *hill_tin* to the scene.

Step 2
Change the TIN's Symbology

In the Table of Contents, double-click *hill_tin* to bring up the Layer Properties dialog.

Click the Symbology tab.

In the Show box, Edge Types is checked. Highlight the words *Edge Types,* and click the Remove button under the Show box.

This prevents the TIN's edges from being symbolized in ArcScene's Table of Contents. Only the Elevation renderer should be left in the Show box, and it should be checked and highlighted.

In the Color Ramp drop-down list, right-click the drop-down arrow and uncheck Graphic View so that you can see the names of the color ramps.

Choose the ramp called Elevation #2. When you're done, the Layer Properties dialog should look like this:

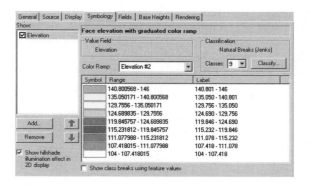

Click OK to dismiss the Layer Properties dialog.

Step 3
Enable Animated Rotation

Click the Navigate tool, and take a look at the hill surface. It's about 42 meters tall.

When you're finished navigating around the TIN, click the Full Extent button.

In the Table of Contents, right-click Scene Layers and choose Scene Properties.

Click the General Tab.

☑ Enable Animated Rotation

When you use the Navigation tool to rotate the scene, hold down the left mouse button, drag in the direction you want the scene to rotate, and release the mouse button while the scene is moving.

Check the box next to Enable Animated Rotation.

Click OK.

Step 4
Start Animated Rotation

Place your cursor at the right side of the display. You'll notice that when animated rotation is active the Navigate tool's cursor is black and white with a circle around it.

Hold down the left mouse button, drag the cursor to the left, and release the mouse button while you drag it. There is an art to this: if you don't release the mouse button until you are well to the left side of the display, ArcScene will treat your command like regular navigation. Try to release the mouse button at the midpoint of the display.

Once the TIN is spinning, take your hand off the mouse.

Step 5
Set Target and Observer During Animated Rotation

If the TIN is rotating too quickly or too slowly, use the Page Up or Page Down key to increase or decrease its speed. (Your cursor has to be in the scene for the Page Up and Page Down buttons to affect the rotation speed.)

The scene rotates when the Navigate tool is active, but you can still use the other tools on the 3D Analyst toolbar. When you click one of them, the rotation is temporarily suspended. To restart the rotation, just click the Navigate tool.

While the TIN is revolving, hold down the right mouse button and zoom in. As soon as the cursor stops moving, the scene resumes its rotation.

Now click the Center on Target tool.

The rotation stops. Click on one of the lower elevation bands on the side of the TIN.

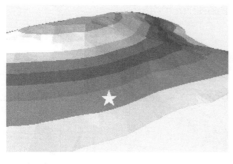

The new target becomes the center of the scene.

Click the Navigate tool again, and the rotation continues, centered around the new target.

Now click the Set Observer tool.

Click on one of the upper elevation bands of the TIN. Try to choose a location that has a direct line of sight to the target.

The scene zooms to the new location and then stays motionless while you look down at the target from the observation point.

Click the Navigate tool again. The scene now rotates 360 degrees around the target, starting from the new observer's position.

Step 6
Stop Rotation

To stop rotation at any time, press the Esc key while the cursor is in the display.

To resume rotation, hold down the left mouse button and drag the cursor to the left or right, releasing the mouse button as you pass the midpoint of the display.

While the TIN revolves, click the Full Extent button. Experiment with some of the other navigational tools.

When you're finished, press the Esc key to stop the rotation. Notice that the Esc key only works when the cursor is in the display.

Step 7
Turn Off Animated Rotation

Double-click Scene Layers in the Table of Contents, and under the General tab uncheck Enable Animated Rotation.

Click OK to dismiss the Scene Properties dialog.

Step 8
Add a Viewer

By adding viewers, you can compare several perspectives of one scene side by side.

Click the Full Extent button.

Click the Add New Viewer button.

A new display appears, titled "Viewer 1."

Move Viewer 1 out of the way of the ArcScene application window.

Click the Navigate tool if it is not already selected, and use it to match Viewer 1 to the following graphic.

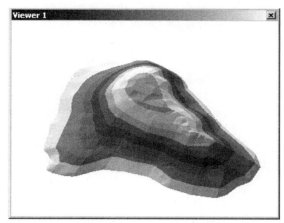

The navigational tools work in the new viewer as well as in the original display. As long as Viewer 1 is active, the tools in Arc-Scene will apply to it.

Step 9
Add Another Viewer

Click the Add New Viewer button again.

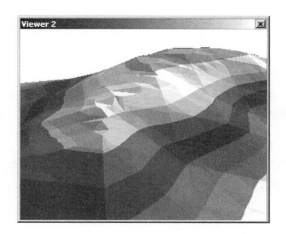

Another display appears, titled "Viewer 2." Notice that whenever you add a new viewer it starts out at the full extent of the data.

Move Viewer 2 so that you can see all three scenes. Use the Navigate tool to steer and zoom in to Viewer 2 until it looks something like the view at left.

Step 10
Apply Scene Properties to All Three Viewers

It might be helpful at this point to distinguish between layer properties, scene properties, and view settings. You know that you can change individual layer properties, such as symbology and base heights, by clicking on a layer in the Table of Contents and bringing up its Layer Properties dialog. Scene properties—illumination, vertical exaggeration, coordinate system, background color—affect all layers equally in a scene, and in fact all viewers of that scene. View settings (such as target and observer positions, orthographic and perspective view, and roll angle and pitch), on the other hand, pertain to navigation and can be tailored for each view separately.

In this step you'll change background color and illumination (both are global scene properties). In the next step you'll change the view settings for two of the scenes.

From the View menu in ArcScene, choose Scene Properties. Move the Scene Properties dialog so that you can see all three viewers.

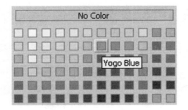

Under the General tab, click the Background color drop-down arrow and choose Yogo Blue.

Click Apply. The sky turns blue in all three viewers.

Now select the Illumination tab.

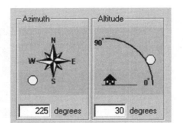

With your cursor, adjust the azimuth and the altitude of the sun. Again, all three viewers reflect the changes.

Type *225* into the azimuth box, and *30* into the altitude box.

Click OK to dismiss the Scene Properties dialog.

Step 11
Change View Settings

From the View menu, choose View Settings. Move the View Settings dialog so that you can see all three viewers.

Click the drop-down arrow next to *Applies to* at the top of the View Settings dialog.

Settings are applied to one viewer at a time. Choose Viewer 1.

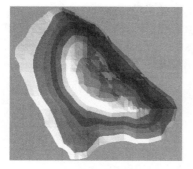

In the *Viewing characteristics* panel, choose Orthographic. This gives you a good top-down view of the hill in 2D. Notice that the Observer's z position and the Target's x, y, and z positions are grayed out.

Now choose Main Viewer from the *Applies to* drop-down list.

In the View Settings dialog, change the Roll Angle from 0 to −35.

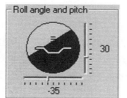

Close the View Settings dialog box.

Now double-click Scene Layers in the Table of Contents to bring up the Scene Properties dialog.

Click the General tab, and check Enable Animated Rotation.

Click OK.

Make sure the Navigate tool is selected. The cursor changes to let you know that animated rotation is active.

Place the cursor over the Main Viewer, and then hold down the left mouse button and swipe it either direction across the scene, releasing it somewhere toward the middle.

Use the Page Up and Page Down keys to speed up or slow down the rotation.

Now hold down the left mouse button and swipe the cursor over Viewer 2. If you don't have too many applications open and you have a powerful computer, you can set it spinning as well.

Press the Esc key to stop the rotation in Viewer 2. Put your cursor in the Main Viewer, and press the Esc key to stop its rotation.

You may notice a couple of things: (1) if you put your cursor over Viewer 1, it looks like animated rotation is available, but Viewer 1 is in orthographic perspective and thus rotation in that viewer won't work, and (2) even though Animated Rotation in enabled from the Scene Properties dialog, it is not a global property for all scenes. Since you activate it with the mouse, you can apply it individually to any viewer.

Step 12
Manage Viewers

From the Window menu, choose Viewer Manager.

The Viewer Manager lets you rename, show, hide, restore, or close any viewers other than the main ArcScene application. If you were to click Close Viewer(s) at any time, you would delete the viewer that is highlighted. Likewise, if you were to click the X in the upper-right corner of a viewer, you would delete it. The Restore option applies to a viewer that has been minimized or maximized, not closed. In this case, we want to keep our viewers, so don't click those options.

With Viewer 1 highlighted, left-click once on the name. A text box with a blinking cursor appears around it, inviting you to type over the name. Do so, and rename it *Ortho_view*.

Press the Enter key to confirm the new name.

Highlight Viewer 2, and rename it *Hilltop_view*.

The two viewers now have more descriptive titles.

With *Ortho_view* selected, click Hide, then Show, then Hide. Keep in mind that this is different from selecting Close Viewer(s).

Click OK to dismiss the Viewer Manager.

Hilltop_view should still be visible. Double-click the middle of its title bar to maximize it.

Now double-click the title bar again to return it to normal size.

Right-click the title bar and take note of the options for creating a bookmark, and for maximizing, minimizing, restoring, and deleting the viewer.

Step 13
Close ArcScene

Experiment as much as you like with View Settings, Scene Properties, and the Viewer Manager. When you're finished, close ArcScene. If you want to save the scene, give it a name of your choice and save it in your *GTK3D\Chapter03\MyData* folder.

EXERCISE 3: THE FLY TOOL

Now that you've had a chance to use animated rotation, you'll test your wings with the Fly tool, which lets you control your speed, altitude, and direction as you navigate through a landscape.

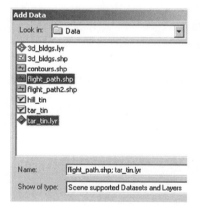

Step 1
Start ArcScene and Add Data

Open ArcScene, and click the Add Data button. Navigate to your *GTK3D\Chapter03\Data* folder.

Add *flight_path.shp* and *tar_tin.lyr* to the scene.

You used *tar_tin.lyr* in Exercise 1 when you learned how to set targets and observers. Since it's fairly small, with both flats and hills, it makes good terrain for flight practice.

Step 2
Symbolize the Flight Path

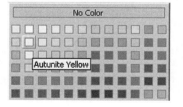

In the Table of Contents, click the line symbol under *flight_path* to bring up the Symbol Selector.

Choose the Highway symbol. Click the Color drop-down arrow in the Options box and choose Autunite Yellow.

Click OK to close the Symbol Selector dialog.

Step 3
Look at the Shapefile's Attribute Table

The flight path was drawn as a 2D polyline shapefile in ArcMap, and then converted to 3D using *tar_tin* as the elevation source.

Right-click *flight_path* and choose Open Attribute Table.

The Z in the Shape field indicates that this is a 3D shapefile. You could also use a 2D shapefile and set its base heights to *tar_tin*, but converting it to 3D saves a step if you share your 3D data or use the same data sets regularly. You'll learn more about 3D shapefiles in Chapter 6.

Close the attribute table.

Step 4
Use the Fly Tool

Fly Tool Instructions	
Action	**Command**
Activate Fly Tool	Click the Fly Tool, then left-click on the scene
Start flight	Left-click
Increase speed	Additional left-clicks
Reduce speed	Right-click
Stop flight	Press the Escape key, or click the middle mouse button

The Fly tool takes a certain amount of practice—more for some, less for others—but in any case you'll need to memorize the commands to use it.

Some other important facts:

You can fly backward by starting your flight with a right-click. Further right-clicks will increase your backward speed; left-clicks will reduce it.

The Shift key is invaluable when using the Fly tool. Holding down the Shift key while flying forward or backward maintains your altitude, and keeps you from crashing into the surface or flying off into space.

To fine-tune your travel speed, press the Up or Down arrow key between left or right clicks of the mouse.

Click the Fly tool.

Put your cursor over the scene. It appears as a bird standing on a cloud until you click it.

Click the left mouse button once, and look down at the status bar.

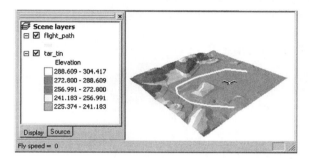

At this point, your fly speed is set to zero, and you can change the direction you're facing before starting the fly-through. Once you're in motion, flight through the scene will follow the movements of the mouse.

Using the commands listed on the Fly Tool Instructions table, practice flying over the surface in various directions and at various speeds. You may want to use the Navigate tool and the Zoom-In tool to get your bird to a good starting position.

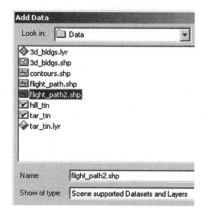

Work on it until you feel somewhat comfortable controlling your flight. Try flying forward along the yellow line, then backward along it. If you lose control, press the Esc key, reorient yourself with the Full Extent button, and try again. Don't worry if you aren't an expert right away; like riding a bicycle, it takes a few sessions.

Step 5
Add Another Flight Path

Click the Add Data button, and add *flight_path2.shp*.

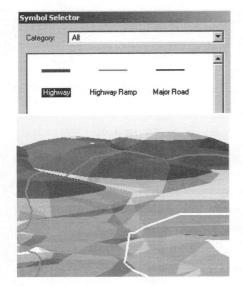

Step 6
Symbolize the New Flight Path

In the Table of Contents, click the line symbol under *flight_path2* to bring up the Symbol Selector.

Choose the Highway symbol. This time leave it bright red.

Click OK to close the Symbol Selector dialog.

Step 7
Fly Through the Mountains

Flight_path2 takes you through hillier terrain. It's a little more interesting, and more difficult to follow.

Use the Navigate tool to get to a comfortable starting position again.

Click the Fly tool, and practice flying through the hills and valleys of the surface. If you can follow the red path all the way to the end, congratulate yourself!

Step 8
Close ArcScene

Experiment as much as you like with the Fly tool. If you want to save the scene, give it a name of your choice and save it in your *GTK3D\Chapter03\MyData* folder.

EXERCISE 4: CREATE 3D ANIMATED FILMS

The animation functions in 3D Analyst let you record, play, save, and share films by manipulating data properties and scene perspectives. The animation capabilities are quite numerous, as you'll see. In this exercise you'll make several short films with data sets that cover a region of the Kentucky River.

Step 1
Start ArcScene and Add Data

Open ArcScene and click the Add Data button.

Navigate to your *GTK3D\Chapter03\Data* folder.

Add *raft.lyr, course.shp, scole_doq.tif,* and *scole_tin* to the scene.

Step 2
Set Base Heights for the Aerial Photograph

Scole_doq.tif is a digital orthophoto—an aerial photo that has been georeferenced and corrected for camera tilt and terrain relief.

In the Table of Contents, right-click *scole_doq.tif* and choose Properties.

Click the Base Heights tab.

Click *Obtain heights for layer from surface*, and choose *scole_tin* from the drop-down list.

Click OK to close the Layer Properties dialog.

Step 3
Lower the TIN's Drawing Priority and Turn Off Edge Types

The TIN and the aerial photo are competing for visual space.

In the Table of Contents, double-click *scole_tin* to bring up its layer properties.

Click the Rendering tab. In the Effects panel, choose a drawing priority of 5 from the drop-down list.

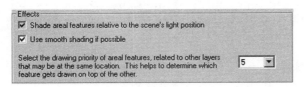

Click Apply.

Now click the Symbology tab. In the Show box, uncheck Edge Types.

Click OK to close the Layer Properties dialog.

The TIN now has a lower drawing priority than the rest of the data, so it doesn't poke up out of the orthophoto.

Collapse the legend for *scole_tin* by clicking the minus sign in the box next to it. This makes it easier to see all the layers in the Table of Contents.

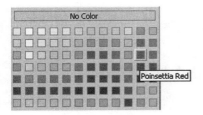

Step 4
Symbolize the Course Layer

In the Table of Contents, right-click the line symbol under *course* to bring up the Color palette.

Change the color to Poinsettia Red.

Your scene should now resemble the graphic at left, with *scole_tin*'s drawing priority set to 5, *scole_doq.tif*'s base heights set to the TIN layer, and *course.shp* symbolized in red.

Step 5
Create Bookmarks

Click the Full Extent button.

From the View menu, click Bookmarks, and then Create.

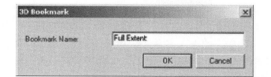

Name the bookmark *Full Extent*.

Click OK.

Use the Navigate and Zoom In tools to arrive at the following perspective:

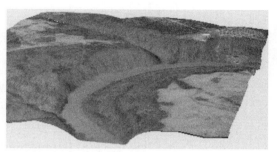

From the View menu, create another bookmark. Name it *River Bend*.

Now zoom to the treatment plant at the northeast corner of the orthophoto.

Create a third bookmark, and name it *Treatment Plant*.

Finally, navigate around to the north side of the treatment plant. Create a fourth bookmark, and name it *Backside*.

Step 6
Load the Animation Toolbar

From the View menu, choose Toolbars, then Animation.

The Animation toolbar lets you create films from a series of captured scene settings, layer properties, or camera positions. It works a lot like traditional animation: each camera track consists of "keyframes" (or snapshots) of the data. The tracks are stored in the ArcScene document, and can be shared as *.asa* (ArcScene Animation) files or as *.avi* files for video playback.

Step 7
Capture Perspective Views to Create Animation

The most basic way to make an animation track is to capture camera positions in a scene and store them as keyframes (snapshots). ArcScene creates the track by interpolating between the keyframes.

Click the Full Extent button again.

On the Animation toolbar, click the Capture View button to make the first keyframe.

Zoom to the southeast corner of the aerial photo. (Remember that at Full Extent you're looking at the surface from the southwest.)

Click the Capture View button again.

Now use the Full Extent, Navigate, and Zoom In tools to head over the northwest corner of the photo.

Click the Capture View button to make the third keyframe.

Navigate and zoom to the treatment plant in the northeast corner, and click the Capture View button a fourth time.

Zoom to the center of the photo, and make a fifth keyframe.

Finish by zooming to the southwest corner and making a final keyframe.

Step 8
Play Back Your Animation

On the Animation toolbar, click the Open Animation Controls button.

The Animation Controls operate just like a video cassette or DVD player, with Play, Pause, Stop, and Record buttons.

Move the Animation Controls out of the way of the scene, and click the Play button.

ArcScene interpolates between the six camera positions you captured, taking a trip from the southeast to the northwest to the northeast to the southwest.

Step 9
Slow Down the Animation

The animation is playing a little too quickly, so you'll change the length of the track.

On the Animation Controls, click the Options button.

In the Duration box, type 7.

Click the Options button again to collapse the dialog box.

Click Play. Now the track takes seven seconds to run instead of three, presenting a slightly more relaxed tour of the terrain.

Step 10
Clear the Animation

From the Animation drop-down menu (not the Animation Controls), choose Clear Animation.

This removes all animation tracks from the scene. If you'd like to start over and make a new track with the Capture View button, you can clear the animation track at any time.

Step 11
Record Real-time Navigation

In addition to interpolating between camera positions, ArcScene can also record navigation in real time. For example, you could make a film in which the data spins around its axis while you zoom in and out to places of interest. In this step, you'll record navigation with the Fly tool.

If you made some more tracks with the Capture View button, choose Clear Animation from the Animation menu.

Click the Full Extent button.

In the Table of Contents, uncheck *scole_doq.tif* so that you can see the TIN underneath.

Click the Fly tool.

On the Animation Controls, click the Record button.

Click the Fly tool once in the scene to get oriented, then again to begin the flight. Navigate around the TIN for as long as you like. When you're finished, press the Esc key.

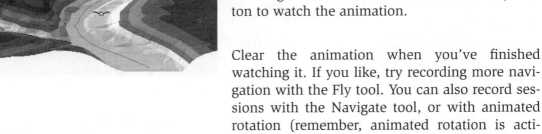

Click the Stop button on the Animation Controls.

Your flight has been recorded. Click the Play button to watch the animation.

Clear the animation when you've finished watching it. If you like, try recording more navigation with the Fly tool. You can also record sessions with the Navigate tool, or with animated rotation (remember, animated rotation is activated from the General tab in the Scene Properties dialog).

Step 12
Use Bookmarks to Make a Camera Track

If necessary, select Clear Animation from the Animation drop-down menu, and then click the Full Extent button.

Click the Navigate button to disable the Fly tool, and turn on *scole_doq.tif* again in the Table of Contents.

Click the Animation menu and choose Create Keyframe.

In the Create Keyframe dialog, choose Camera from the Type drop-down list.

Click New to create a new track.

Leave the name *Camera track 1* and click OK.

In the Create Keyframe Animation dialog, click Create.

You just created the first keyframe. Because Import from Bookmark wasn't checked, Arc-Scene just used the current camera angle in the scene, at full extent.

Now check Import from Bookmark. From the adjacent drop-down menu, choose River Bend.

Click Create.

Now choose Treatment Plant from the drop-down menu, and then click Create.

Choose Backside, and then click Create.

Choose Full Extent, and then click Create one last time.

Click Close.

On the Animation Controls, click the Options button, and type *10* in the Duration box. Click Options again to collapse the dialog.

Click the Play button to run the animation.

Step 13
Add the 3D Effects Toolbar

From the View menu, click Toolbars, and check 3D Effects.

The 3D Effects toolbar contains several tools that affect lighting, shading, drawing priority, and transparency. It also has a tool for face culling, which lets you see through areal features by hiding either their back or front views, depending on your observer's position.

On the 3D Effects toolbar, select *scole_doq.tif* from the Layer drop-down list.

Step 14
Change Layer Properties During Animation

From the Animation menu, choose Create Keyframe.

From the Type drop-down list, choose Layer.

From the Source Object list, choose *scole_doq.tif*.

Click New, and leave the name *Layer track 1*.

Click OK.

The Create Animation Keyframe should look like this:

You're going to use the Layer Transparency tool to fade the aerial photo in and out during animation.

Move the dialog out of the way of the scene.

Both *scole_tin* and *scole_doq.tif* should be turned on in the Table of Contents. Also reassure yourself that *scole_doq.tif* is selected in the 3D Effects toolbar.

In the Create Keyframe dialog, click Create.

On the 3D Effects toolbar, click the Layer Transparency tool.

Slide the transparency control all the way from 0 to 100 percent.

In the Create Keyframe dialog, click Create again.

Now slide the transparency control back to 0 percent.

Click Create three (3) more times. This way, the aerial photo will fade up well before the end of the animation track.

Close the dialog.

Step 15
Play Back the Animation

If you closed the animation controls, click the Open Animation Controls button on the Animation toolbar.

Click the Play button.

Both tracks play at the same time: the Camera track (which interpolates between the various bookmarks you made) and the Layer track, which fades the aerial photo out and back in again.

Step 16
Open the Animation Manager

The Animation Manager lets you fine-tune your animation in a variety of ways. For example, you can add, delete, enable, and disable tracks; toggle keyframes; adjust the location of keyframes within track play; and change the percentage of total play time each track fills. To learn more about the Animation Manager's capabilities, see Chapter 8 in the *Using ArcGIS 3D Analyst* manual. (In ArcGIS 8.2 and 8.3, Chapter 8 and its companion, 3D Analyst Tutorial Exercise 5, are *.pdf* files bundled with the ArcGIS installation software.)

From the Animation menu, choose Animation Manager.

Click the Tracks tab.

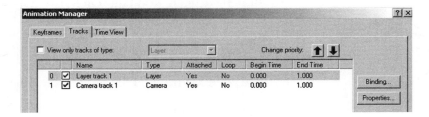

Take a look at the tracks. Both the Camera track and the Layer track are enabled, and they start and end together.

Click the Time View tab.

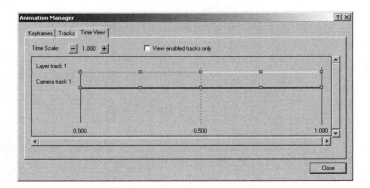

The Camera track and the Layer track happen to have five keyframes each, and in this case they punctuate the track runtime at the same points. The Animation Manager treats keyframes as events that happen along a percentage of the entire play time. (To alter the number of real seconds an animation plays, as you may remember from Step 9, you use the Options button in the Animation Controls.)

Step 17
Preview the Animation at Specific Moments Along the Track

Move the Animation Manager as well as you can away from the scene.

In the Animation Manager, click the cursor anywhere within the timeline. Try to click in negative gray space, rather than on one of the horizontal track lines. A vertical red line appears, and its relative position on the timeline is listed in red at the lower left of the Time View panel.

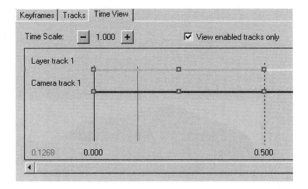

Look at the scene. It shows you what the animation looks like at that point in the track.

Click a few other places along the timeline, and watch the scene.

When you're finished, click anywhere outside the timeline.

Step 18
Set the Layer Track to Play Before the Camera Track

The keyframes along each track are represented by small squares. You can move these wherever you like, and the track will play accordingly.

Using your cursor, grab the squares and move them, one by one, along the Layer track and the Camera track until their placement resembles the graphic below:

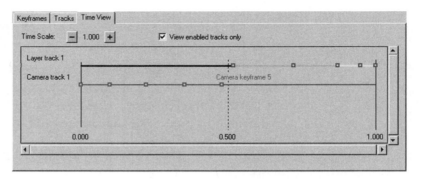

You've set up the animation so that the Camera track runs during the first half of the timeline, and the Layer track runs during the second half.

Click Close.

Step 19
Play the Animation

On the Animation Controls, click the Play button.

This time, the camera makes a path through the bookmarked locations, and then the aerial photo fades out and back in. Notice that each track takes about half the amount of time it did before.

Feel free to experiment with the Time View settings and the two tracks.

Step 20
Remove the Layer Track and Reset the Camera Track

From the Animation menu, open the Animation Manager. Click the Tracks tab.

Select the Layer track, and click Remove. Only the Camera track is left.

Click the Keyframes tab.

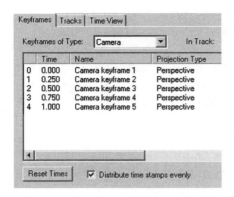

Make sure the box next to *Distribute time stamps evenly* is checked, and then click Reset Times.

Click the Time View tab, and then look at the Camera track. The keyframes are evenly distributed over the timeline again.

Close the Animation Manager, and then click Play to run the Camera track again.

Step 21
Clear all Animation

From the Animation menu, choose Clear Animation.

Step 22
Move a Layer Along a Path

The expression "move a layer along a path" may sound somewhat mysterious, but it means just that: you can move the midpoint of any geographic layer along a selected line feature or line graphic in the scene. In this case, you're going to move the *raft* layer along the *course* layer.

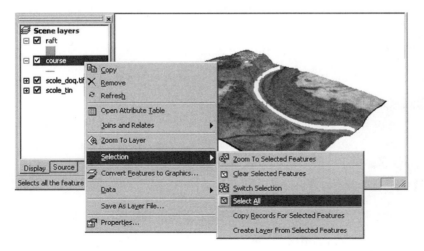

In the Table of Contents, right-click *course*, choose Selection, and click Select All.

From the Animation menu, choose Move Layer Along Path.

From the Layer drop-down menu, choose *raft*.

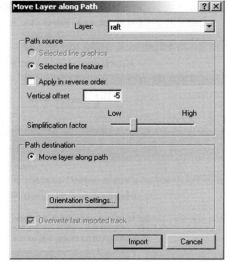

The Path Source is the line feature (*course.shp*) you selected in the scene.

In the Vertical Offset box, type -5. This submerges the raft somewhat so that it looks less like a lumpy brown racecar.

Set the Simplification factor about a quarter of the way along the bar, if it isn't there already. The higher the simplification, the smoother the animation, but the more it taxes your system.

Click Import.

In the Table of Contents, uncheck *course*. (The *line* shapefile is still selected, just not visible.)

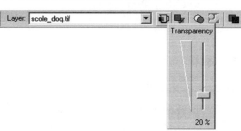

In the 3D Effects toolbar, select *scole_doq.tif* from the drop-down menu. Click the Transparency tool, and set the aerial photo's transparency to about 20 percent.

If the Animation Controls aren't active, click the Animation Controls button on the Animation toolbar.

Click the Options button on the Animation Controls.

In the Duration box, type *10*.

Click Options again to close the panel.

Step 23
Play the New Track

Click the Play button.

The raft traces a course along the bend in the Kentucky River.

Step 24
Rename the Track, and Disable It

From the Animation menu, choose Animation Manager.

Click the Tracks tab.

Click the words *Track from path*. A box appears around the high-lighted name. Rename the track *raft*, and then press the Enter key.

Now uncheck the box next to the *raft* track. You'll use it again later, but for the next step it should be disabled.

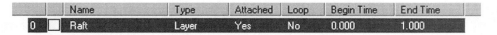

		Name	Type	Attached	Loop	Begin Time	End Time
0	☐	Raft	Layer	Yes	No	0.000	1.000

Close the Animation Manager.

Step 25
Create a Camera Flyby from a Path

You just used the *course* shapefile to set a path for the *raft* layer to follow. You can also use that line feature to create a path for a target to move along. In this step, you'll have an observer (the camera) sit in one location and turn to watch the target move along the path.

Zoom and navigate to the outer bank of the bend in the river, like so:

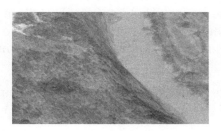

From the Animation menu, choose Camera Flyby from Path.

The path source is the *course* shapefile. It should still be selected, but turned off, in the Table of Contents.

For the path destination, choose the last option, *Move target along path with current observer*. This means that the current position of the camera—zoomed in on the bank of the river—stays still while the target moves along the path.

Click Import.

Click the Play button on the Animation Controls.

ArcScene moves the (invisible) target along the path you specified. The camera's location doesn't change, but it does turn to
keep the target in view.

Step 26
Play Both Tracks Together

Open the Animation Manager, and click the Tracks tab.

Turn on the *raft* track. Leave the new Track from Path on as well.

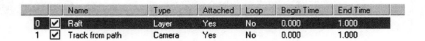

Click Close.

Click the Play button.

The tracks play simultaneously: the *raft* layer travels along the
course layer, and the camera observes a target that traverses the
same path.

Step 27
Close ArcScene

This exercise covered only a few of ArcScene's animation functions. See the 3D Analyst Help and Exercise 5 in the *Using 3D
Analyst* manual to learn more. In the meantime, feel free to experiment with the Animation tools.

When you're finished, close ArcScene. If you want to save your
changes, name the scene and save it in your *GTK3D\Chapter03\
MyData* folder.

CHAPTER 4

RASTER SURFACE MODELS

RASTER INTERPOLATION

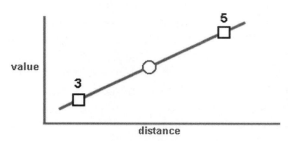

An example of linear interpolation. The center dot lies midway between two known values, 3 and 5, so its value is estimated to be 4.

In Chapter 1 you learned about the cell structure of raster data. You also learned that in a thematic raster samples of the phenomenon are used to estimate the unknown cell values in the model. This estimation process is called interpolation, and is used to create models of measurable phenomena, such as elevation, rainfall, air pollution, snow depth, noise, and contamination.

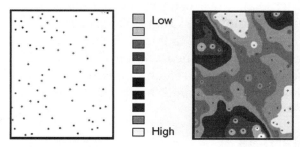

An elevation grid created from height samples.

As you might imagine, interpolating values across a surface is more complicated than estimating them along a straight line. In the grid at left, an elevation surface has been interpolated from a set of points. Each point represents a location where the elevation has been measured. The resulting grid is a prediction of what the elevation is at any location on the actual surface. The more points accurately sampled, the better the prediction.

113

Points on a surface show where the samples were taken. Extra samples were taken where the slope changes rapidly, in order to create a more accurate model.

Numbered cells indicate known values. In estimating the value of the unknown center cell, an interpolation method using spatial autocorrelation will use all six sample points, but will give more influence to the nearest ones. Here, the darker cells indicate the most influential sample points.

Spatial interpolation can be used to create a surface from just a few sample points. However, more sample points are better if you want a detailed model. Input points can be randomly, deliberately, or regularly sampled. For an area where the surface changes sharply or the phenomenon is concentrated, a cluster of sample points may be necessary.

The central notion behind spatial interpolation is that points near each other are more alike than those further apart. Therefore, the value at any location should be estimated using nearby points. Most interpolation methods apply this principle by giving the nearest sample points the most influence when estimating an unknown cell value. 3D Analyst offers several interpolation methods for creating raster surfaces from point data, including Inverse Distance Weighted (IDW), Spline, Natural Neighbors, and Kriging. (The Trend interpolation method is offered as well, but you need to customize 3D Analyst in order to use it.)

Sample Size

Most interpolation methods let you control the number of sample points used to estimate cell values. If you limit your sample to seven points, the interpolator will use the seven nearest known values to estimate each unknown cell value. You can also control the sample size by defining a search radius. A fixed search radius will use only those samples contained within it. If not enough sample points are found within the search radius, you can use a variable search radius that expands until the specified sample size is found.

Interpolation Barriers

What if your surface contains a sharp break in terrain, such as a river bank or a cliff. If you interpolate values on either side, the sharp break will be smoothed over. One interpolation method (IDW) lets you include barriers in your analysis. The barrier works like a force field, essentially interpolating two separate surfaces. Unknown values on one side do not take any samples from the other side for analysis. Interpolation barriers can add considerably to the time it takes to compute unknown values. One way to speed up the process is to use as few vertices in the barrier feature as possible.

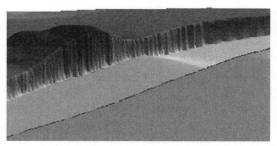

If terrain contains a sudden change in elevation, you can set an interpolation barrier to keep samples at one side of the break from affecting values on the other side of the break.

IDW, Spline, Kriging, and Natural Neighbors interpolation methods work differently, and which method will produce the most accurate surface will depend on the phenomenon you're modeling and the distribution of sample points. No matter which method you use, the more evenly distributed your input points the more accurate the results.

Inverse Distance Weighted (IDW) Interpolation

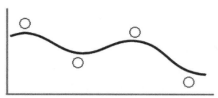

Because IDW averages values, a surface created with it won't pass through or exceed the value range of the sample points.

IDW is the least complicated interpolation method. Unknown cell values are calculated by averaging the values of sample points in the vicinity of each cell. The closer a sample point is to the center of the cell being estimated, the more influence it has in averaging the cell's value. A specified number of points, or all of the points within a given radius, can be used to compute the value of each output cell.

IDW works best for closely packed, consistently spaced sample point sets. It tends to flatten out surface variation rather than preserve it. It's good for measuring phenomena such as sound, whose distribution falls off sharply with distance. It's not so good for measuring phenomena such as air pollution, because it can't take into account any prevailing trends in the data, such as wind. Because IDW averages values, the surface model won't pass through the sample points, and no estimates will be made above the maximum or below the minimum sample values.

That being said, IDW offers a number of adjustable parameters other interpolation methods don't. IDW comes with a Power setting that adjusts the relative influence of the sample points. A Power setting of 0 gives all samples (within the specified search radius) almost equal influence in estimating the output point. As the Power setting increases, the influence of sample points falls off more rapidly with distance. The output cell values are more localized, so the new surface has more detail. A power of 2 is most commonly used.

You can also choose the type of radius used to control the number of sample points. A Fixed Radius setting uses a constant distance, including all of the sample points found within that radius, to interpolate an unknown cell value. A Variable Radius setting lets you specify the number of points that will be averaged. The radius for each interpolated cell will be different, because it has to expand until it finds the specified number of sample points. It's better to use a variable radius when the density of sample points varies significantly from one area to another. If you have areas with sparsely distributed input points, you can place a limit on the radius, so that if the number of points isn't reached inside the maximum distance of the radius the interpolator will just use fewer points in the calculation.

IDW also lets you use polylines to limit the search for sample points. The lines may represent a cliff, a river, or some other interruption in a landscape. As discussed earlier, only sample points on the same side of the barrier as the unknown cell will be used to estimate it.

Spline Interpolation

Instead of averaging values between sample points, the Spline interpolator creates a surface that passes exactly through them. This is useful if you want to be able to estimate values that are below the minimum or above the maximum values found in the sample data.

Spline interpolation is best for surfaces that vary gradually. If samples are close together and have extreme differences in value, it can overshoot estimated values because it uses slope calculations (change over distance) to figure out the shape of the surface. Terrain or phenomena that change suddenly, such as a cliff face or a fault line, are not well represented by a smooth curving surface. In those cases, you might be better off using IDW interpolation.

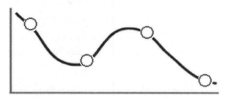

The Spline interpolator creates a surface that passes through each sample point. The surface values may surpass the sample values.

There are two Spline methods, Regularized and Tension. Regularized creates a smooth, gradually changing surface with values that may lie outside the sample data range. Tension forces the estimates to stay closer to the sample data, resulting in a tighter, less elastic surface.

Kriging Interpolation

Kriging is similar to IDW in that it weights nearby point samples to estimate an unknown value. In IDW, however, this weight depends solely on the distance to the unknown location. Kriging assumes that the distance and direction between sample points indicate a spatial relationship in the surface, so the weights are based not only on the distance between the measured points and the unknown location but on the overall arrangement among the measured points. It adapts its calculations to the data by analyzing all of the data points to find out how much correlation they exhibit, and then factors it in the weighted average estimation.

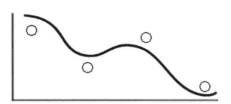

A surface created with kriging can exceed the value range of the sample points but won't pass through the points.

Kriging involves several steps before it creates the surface, including exploratory statistical analysis of the data and variogram modeling. It's a very complicated interpolation engine, that's most appropriate when you know there is a distance or directional bias in your data. It's often used in soil science and geology. A great deal more information can be found in the ArcGIS 3D Analyst Help, and in the 3D Analyst user manual.

Two commonly used kriging methods are ordinary and universal kriging. Ordinary kriging assumes that there is no particular trend in the data, whereas universal kriging assumes that there is a prevailing movement across the surface. For example, you may know that wind is coming from one direction, or that the terrain slopes consistently. Universal kriging applies a polynomial function to the area. The polynomial is subtracted from the original sample points, and the autocorrelation is estimated from the random errors. Once the model is fit to the random errors, the polynomial is added back before making an estimation of the surface. Universal kriging should be used only when you know there is a prevailing trend in your data.

Natural Neighbors Interpolation

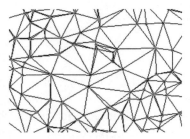

Delaunay triangulation connects irregularly located points with lines to form a network of contiguous, nonoverlapping triangles.

Like IDW, Natural Neighbors interpolation is a weighted-average method. However, instead of estimating a point's value using sample cells weighted by their distance, the Natural Neighbors interpolator first performs a Delaunay triangulation. A Delaunay triangulation is what is used to create a TIN from sample points. We'll cover that in more detail when we talk about TINs, but suffice it to say that a Delaunay triangulation connects all sample points so that each point is a node in a triangle, and all triangles are contiguous and nonoverlapping.

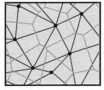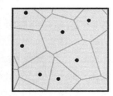

On the left is a set of sample points connected by Delaunay triangulation, with the Voronoi polygons in the background. On the right, the Voronoi polygons are drawn by bisecting each triangle side. The result is that each polygon encloses the area that is closer to a point than to any other point.

When the triangles are formed, Natural Neighbors creates a set of Voronoi polygons around the sample points. Voronoi polygons, also called Thiessen polygons, are formed by drawing the perpendicular bisector of each of the triangle lines so that each Voronoi polygon bounds the region that is closer to one point than to any other point.

Finally, the raster surface is interpolated using the sample data points that are natural neighbors of the cell centers. The value of an estimated location is a weighted average of the values of the natural neighbors. Since the output is a raster, the estimation locations are a regularly spaced array equal to the number of raster cells.

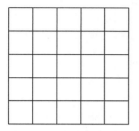 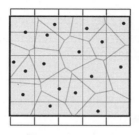

For each cell in the output raster, a weighted average of the cell center's natural neighbors is calculated.

Natural Neighbors is most suitable for sample data points that are unevenly distributed. Because it uses a weighted average, it doesn't require specific parameters such as radius, number of sample points, or weights. Remember, when you use any of these interpolation methods you should have a clear understanding of your data and of the phenomenon being modeled.

RASTER RECLASSIFICATION METHODS: SLOPE, ASPECT, HILLSHADE, AND VIEWSHED

Now that you know a bit about how rasters are interpolated from sample points of elevation, let's look at some common ways cell values are reclassified. When you reclassify a raster, you replace an existing set of cell values with a new (and usually smaller) set of values. 3D Analyst provides engines for calculating commonly desired raster surfaces from elevation values that reveal the slope, aspect, hillshading, and viewsheds of an area.

For rasters of phenomena other than elevation, you may want to assign new values in order combine it with other rasters for further analysis, to change classification schemes, or to assign standards of preference, priority, or sensitivity. Reclassification should not be confused with resampling, which transforms the cell size or coordinate space of a raster. Reclassifying changes cell values but not the resolution of the raster (the number of cells per square unit of area in the grid).

Slope

Slope is the steepness of a surface, measured as the change in surface value over distance. Slope can be measured in degrees (0 to 90) or as a percentage, which is the rise divided by the run times 100. A slope of 45 degrees equals 100-percent slope. Measures of slope in degrees approach 90 (vertical), and measures of slope as percentages approach infinity.

In a raster, slope is calculated as the maximum rate of change in value between a cell and its eight closest neighbors. Areas of low slope are good for construction; areas with high slopes are prone to erosion and landslides.

Aspect

Aspect is the compass direction a slope faces. Values are measured clockwise in degrees from 0 (due north) to 360 (again due north). A slope with an aspect value of 270, for example, faces west. If you walked down that slope, you'd be walking west. (If you walked up the slope, you'd be walking east, so aspect describes the downslope direction a cell faces.) Flat areas get an aspect value of –1.

In a raster, aspect is figured for each cell by estimating the fit of a plane to the z values of that cell plus its eight neighbors. The aspect of the plane becomes the aspect value of the cell in a new raster. Aspect modeling is used in many applications, such as determining the amount of solar heating a building will receive, how much sun vegetation will get, and how fast snow will melt.

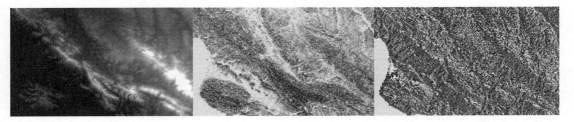

The center and right panels are slope and aspect rasters calculated from the elevation raster on the left. Aspect is usually symbolized with a bright, primary color scheme: one color for each direction, plus gray for flat areas.

Hillshade

Hillshade is the pattern of light and dark a surface shows when lit from a particular angle. It's used to model the amount of sun and shadow an area receives at different times of day, and to increase the perception of terrain relief in a 2D surface.

The Hillshade function in 3D Analyst is like slope and aspect, in that it calculates a new raster based on the sun angle and steepness of each cell on the surface. The value of each hillshaded cell ranges from 0 (no light) to 255 (brightest light).

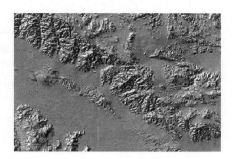

A hillshade raster calculated with the sun's azimuth at 315 degrees (shining from the northwest) and an altitude of 45 degrees above the horizon.

Viewshed

Viewshed analysis identifies the areas of a surface that can be seen from one or more observation points. It is used in applications ranging from the valuation of real estate to the placement of telecommunications towers.

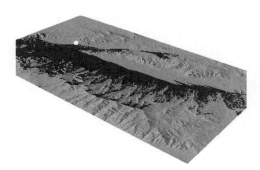

A viewshed raster derived from a shapefile (the white observation point) and an elevation raster. Black cells are visible from the observation location, gray cells are not. The viewshed raster is partially transparent, and a hillshade of the elevation raster is displayed underneath to create a more realistic landscape.

The observation points can be features in a point layer, or the vertices in a line layer. Each cell in the viewshed raster stores the number of observers who can see it. If you have only one observation point, each cell that can be seen from that point gets a value of 1. All cells that can't be seen from that point get a value of 0.

If more than one observer point is used, each visible cell in the raster stores the number of points that can see it. Viewshed analysis assumes that the observer's field of view has no horizontal or vertical constraints, and no distance limits.

More Reclassification

Slope, aspect, hillshade, and viewshed are the rasters most commonly derived from height values. With 3D Analyst, you can use an elevation raster or a TIN as the source for those heights.

Rasters—whether they contain elevation values or not—can be reclassified for a myriad of analysis applications. You may want to replace cell values with updated information, give certain cells a *NoData* value, or classify values together before converting the raster to a vector format. Often, several rasters are reclassified to a common scale so that they can be combined to model a larger spatial problem.

For example, a raster of soil types might be reclassified to create a raster of soil pH values; values in a slope raster might be reclassified into values of suitability; or a raster of wildlife habitat might be reclassified into values of sensitivity or preference. For more information on raster reclassification and its uses, see *About Analyzing Surfaces* and *Reclassifying Your Data* in the ArcGIS Desktop Help.

In Chapter 1 you learned that a difference is often struck between image rasters and thematic rasters. Thematic rasters, in turn, are often divided into a couple of categories: those that model categorical data and those that model continuous data. Although the differences between categorical and continuous phenomena can sometimes be argued, continuous data generally represents a phenomenon that varies continually over a surface, such as elevation, precipitation, or pollution concentration. Categorical data—also called discrete data—represents a phenomenon with known boundaries that can be placed into classes, such as a raster of land-use categories.

ObjectID	Value	Count
0	1	32386
1	2	62737
2	3	84318
3	4	107716
4	5	103833
5	6	97178
6	7	64159
7	8	33833
8	9	13236
9	10	3466

Cell values in this integer raster are classed into 10 categories of buffer distance. The buffers could also be represented by vector polygons.

When you reclassify continuous data, you replace ranges of values with individual values. So, a continuous elevation raster, for example, could be reclassified into two groups: within a flood plain and outside the flood plain.

A continuous elevation raster reclassified into a raster containing two flood plain categories.

When you reclassify categorical data, cell values are replaced on a one-to-one basis. For example, in a soil habitat analysis you might want to replace soil types, represented by numeric codes, with soil pH values. You can also reclassify discrete data into ranges. For example, if you were investigating places to build a new park, you could reclassify a raster that contains 10 land-use categories to create a raster containing just the three preference categories low, medium, and high.

Now that you've been introduced to raster interpolation and classification methods, it's time to create some raster surface models; calculate slope, hillshade, and aspect; and create viewsheds from observer points.

EXERCISE 1: INTERPOLATE A TERRAIN SURFACE WITH SPLINE

In this exercise you'll use the Spline interpolator to create elevation models of San Luis Obispo County, CA.

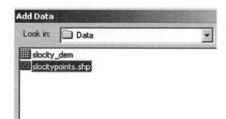

Step 1
Start ArcScene and Add Data

Open ArcScene and click the Add Data button.

Navigate to your *GTK3D\Chapter04\Data* folder.

Add *slocitypoints.shp* to the scene.

Step 2
Examine the Shapefile

Zoom in to the layer and have a look. *Slocitypoints* is a point shapefile of elevation samples in a portion of San Luis Obispo County.

Step 3
Change the Scene to Orthographic View

From the View menu, choose View Settings, and then click Orthographic (2D) View. This will make it easier to compare differences in the raster surfaces you are about to create.

Close the View Settings dialog.

FID	Shape*	POINTID	GRID_CODE
403	Point	3753	24
393	Point	3677	25
404	Point	3754	29
357	Point	3525	30
406	Point	3757	30
342	Point	3451	32
376	Point	3600	32
392	Point	3676	32
325	Point	3372	33
275	Point	3146	34
310	Point	3296	34
341	Point	3447	34
343	Point	3452	34

Record: 14 4 | 0 ▶ ▶I | Show: All Selected | Records (0 out of 407 Selected.)

Step 4
Open the Point Layer's Attribute Table

In the Table of Contents, right-click *slocitypoints* and choose Open Attribute Table.

Right-click the *GRID_CODE* field and choose Sort Ascending.

The attribute table contains 407 records. Scroll through it and look at the sample elevation values in the *GRID_CODE* field. They range from 24 to 366 feet. This data set is a subset of a larger point shapefile; the actual sample values were taken over a larger area and ranged further in value, and these samples were selected at random from the larger set. Some of the highest and lowest values across the area, therefore, are not represented.

Close the attribute table.

Step 5
Set the Analysis Environment

From the 3D Analyst menu, choose Options.

Click the General tab.

Use the Browse button to set the Working directory to *GTK3D\ Chapter04\MyData*, and set the Analysis mask to None.

The Working directory is where analysis results are stored. Most rasters created in 3D Analyst are temporary—that is, they aren't saved to disk. This includes rasters created through interpolation, reclassification, and surface analysis. Exceptions to this rule are the Natural Neighbors interpolator and the TIN-to-Raster conversion function. Rasters created with these methods are permanently saved to a directory you specify.

If you want to make a raster permanent, you can change its temporary status when you create it, saving it either in the working directory or in some other location. You can also right-click a temporary raster in the Table of Contents and choose Make Permanent, which will prompt you to choose a directory. Finally, if you save a map or scene document, any temporary rasters it contains will be saved permanently to your Working directory. Rasters saved in this way are given a name followed by a number. The name describes the type of raster, and the number identifies the raster individually; for example, *IDW8*, *SPLINE5*, or *RCLASS3*.

On the Extent tab, set the analysis extent to *Same as Layer slocity-points*.

On the Cell Size tab, choose As Specified Below from the *Analysis cell size* menu, and type *98* into the cell size box.

Click OK.

Step 6
Interpolate a Surface with Spline

You'll use the Spline interpolator because it can estimate values above and below the range of sample points. There are no abrupt elevation changes, either, so Spline will work well for this surface.

From the 3D Analyst menu, choose *Interpolate to raster*, and click Spline.

Set the Z value field to *GRID_CODE*.

The Output cell size should be 98, and the Spline type should be Regularized.

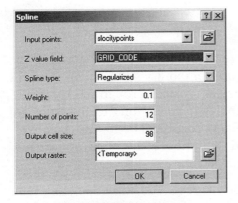

Leave the other parameters at their default settings. The higher the Weight setting in Regularized Spline, the smoother and more rubbery the surface. Values for this parameter have to be equal to or greater than 0. Typical values are 0, .001, .01, .1, and .5. The default value is .1.

Click OK.

The interpolator builds an elevation model based on the values in the *GRID_CODE* field.

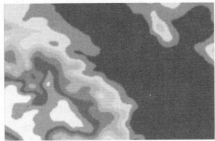

Step 7
Examine the New Raster

In the Table of Contents, notice that the elevation values range from about 18 to 371 feet. This range is slightly greater that the span of values you saw in the attribute table of *slocitypoints*.

Click the name *Spline of slocitypoints* to select it, and then click it again to make a highlighted box around the name. Rename it *Regular Spline 0.1*.

Turn off Regular Spline 0.1 in the Table of Contents, and collapse its legend.

Step 8
Change the Weight and Run Spline Again

You're going to increase the weight setting, amplifying the bend of the surface between the sample points and creating a looser, more flexible interpolation.

From the 3D Analyst menu, choose *Interpolate to raster*, and click Spline.

Make sure *Z value field* is set to *GRID_CODE*.

Change the Weight setting from 0.1 to 1, and click OK.

A second layer, *Spline of slocitypoints*, is added to the scene.

Compare the two surfaces by turning the layers on and off. The differences are slight, but you can see changes, particularly in the lower elevations symbolized in dark green.

Rename the new surface *Regular Spline 1* and collapse its legend. Turn off all layers.

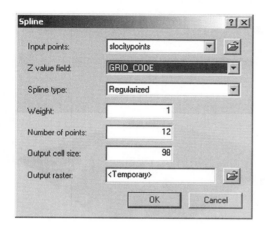

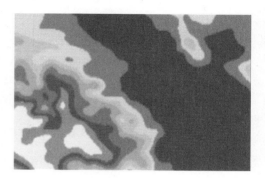

Step 9
Run Tension Spline

The Tension spline setting creates a less elastic surface than Regularized spline, keeping the minimum and maximum estimations closer to the sample data points. The tension curve is flatter than the regularized curve, meaning that there is less overall variation in the final values.

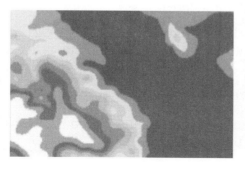

From the 3D Analyst menu, choose Interpolate to Raster, and click Spline.

Make sure *Z value field* is set to *GRID_CODE*.

Change the Spline type to Tension, but leave the rest of the parameters at their defaults.

Click OK.

A third layer, *Spline of slocitypoints*, is added to the scene.

Turn the layers on and off to compare the three surfaces. (The layers draw in the reverse order of their order in the Table of Contents, although if you were to save the ArcScene document, close it, and reopen it they would draw in the conventional drawing order.) The values in the Tension spline correspond more closely to the original range of sample values—24 to 366—in the *slocity-points* shapefile. The *Regular Spline 0.1* surface falls between the tighter Tension spline and the more elastic Regular Spline 1.

Rename the new surface *Tension Spline 0.1* and collapse its legend.

Step 10
Set Layer Properties

From the View menu, click View Settings. Change the projection from Orthographic to Perspective.

Close the dialog.

In the Table of Contents, turn on all layers.

Right-click Tension Spline 0.1 and choose Properties.

Click the Source tab, and look at the Data Source panel. The temporary rasters you've been creating have the name *SPLINE* followed by a number, and are saved in the Working directory you specified when you set up the analysis environment in step 5. If you made more than three rasters during this exercise, your raster may have a higher number, but in any case the most recent raster you created will have the highest number.

```
Data Type: Raster
Folder: C:\GTK3D\Chapter04\MyData\
Raster: SPLINE3
Status : Temporary

Coordinate System:
```

Now click the Base Heights tab. Click *Obtain heights for layer from surface*, and make sure the appropriate raster is selected.

In the Offset panel, type *1500*.

On the Rendering tab, click to shade areal features, and then click OK to close the Layer Properties dialog.

Open the layer properties for the other two layers. Set their base heights to the appropriate rasters and shade their areal features. For Regular Spline 1, give it an offset of 750. Leave the offset for Regular Spline 0.1 set to 0.

Step 11
Set Scene Properties

Now you have a nice sandwich of surface rasters, and you can tell which one you're looking at when you turn them on and off. But let's exaggerate the vertical to make the surfaces a little more interesting.

In the Table of Contents, double-click Scene Layers to bring up Scene Properties.

Click the General tab, set the Vertical Exaggeration to 5, and click OK.

Zoom to the full extent of the scene and turn on all three spline surfaces.

Step 12
Navigate the Scene and Save the Document

Zoom in closely to the three surfaces, turning layers on and off. When you're finished, save the ArcScene document under a name of your choice in your *GTK3D\Chapter04\MyData* folder. Then close ArcScene.

EXERCISE 2: INTERPOLATE TERRAIN WITH INVERSE DISTANCE WEIGHTED AND NATURAL NEIGHBORS

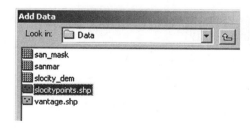

Step 1
Start ArcScene and Add Data

Open ArcScene and click the Add Data button.

Navigate to your *GTK3D\Chapter04\Data* folder.

Add *slocitypoints.shp* to the scene.

Step 2
Set the Analysis Environment

From the 3D Analyst menu, choose Options.

On the General tab, set the Working directory to *GTK3D\ Chapter04\ MyData*.

Set the Analysis mask to None.

On the Extent tab, set the analysis extent to *Same as Layer slocity-points*.

On the Cell Size tab, choose As Specified Below and set the cell size to 98.

Click OK.

Step 3
Change the Scene to Orthographic View

From the View menu, choose View Settings, and click Ortho-graphic (2D) View.

Close the View Settings dialog.

Step 4
Interpolate a Surface with IDW

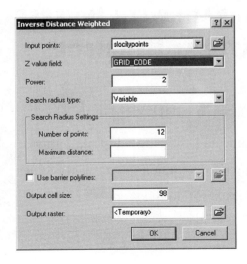

From the 3D Analyst menu, choose Interpolate to Raster, and click Inverse Distance Weighted.

Set the Z value field to *GRID_CODE*.

The Output cell size should be 98, and the Power setting should be 2.

Leave the other parameters at their default settings, and click OK.

A new elevation model, *IDW of slocitypoints*, is added to the scene.

Step 5
Set Base Heights for slocitypoints

In the Table of Contents, turn on *slocitypoints*, and double-click it to bring up its layer properties.

On the Base Heights tab, click *Obtain heights for layer from surface*, and choose the new *IDW* raster from your *Chapter04\ MyData* folder.

Click OK.

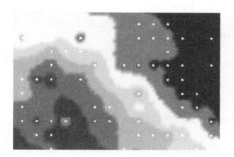

In the last exercise, you studied three surfaces of this area created with the Spline interpolator. There are a couple of things to notice about the IDW surface: (1) the range of interpolated cell values is within the range of values from the original point shapefile, because IDW averages values, and (2) the IDW interpolator has left rings around the sample points, like little grease spots.

The sample points are exerting too much local influence because the Power setting is too high. A Power setting of 0 gives near and far points almost equal influence in estimating a cell's output value. As the Power setting increases, the influence of nearby sample points is given greater weight. The output cell values become more localized and less averaged, and the resulting surface will have more detail. In this case, the detail is exaggerated.

In the Table of Contents, click *IDW of slocitypoints* once to select it, and then again to make a highlighted box around the name. Rename it *IDW Power 2*.

Turn off IDW Power 2 in the Table of Contents, and collapse its legend.

Step 6
Change the Power Setting and Run IDW Again

From the 3D Analyst menu, choose Interpolate to Raster, and click Inverse Distance Weighted.

Make sure the Z value field is set to *GRID_CODE*.

This time, change the Power setting to 0.4.

The Output cell size should be 98, and other parameters should be at their default settings.

Click OK.

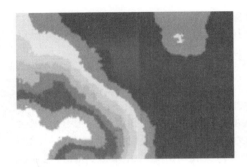

A new elevation surface is added to the scene. It looks quite different from the *IDW* surface that had a Power setting of 2. The local influence of the sample points has been decreased, resulting in a much more generalized surface.

Rename the surface *IDW Power 0.4*, and collapse its legend.

Turn off all layers in the Table of Contents.

Step 7
Interpolate the Surface with Natural Neighbors

Since the Natural Neighbors interpolator is designed for irregularly sampled values, it should work well for *slocitypoints*—some areas of which are sampled densely and some areas of which have no samples.

From the 3D Analyst menu, choose Interpolate to Raster, and click Natural Neighbors.

The Natural Neighbors interpolator doesn't have many parameters; that is, you can't specify a radius, a weight, or a number of sample points to be used. And unlike IDW and Spline, it's not affected by most of the Analysis environment parameters you set in the Options dialog. For example, you can't control the extent of the output raster, and you can't use a mask.

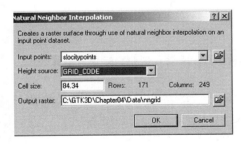

Make sure the Height Source field is set to *GRID_CODE*.

Let the Natural Neighbors dialog establish the rest of the parameters.

Click OK.

Wow! The surface looks pretty weird. Unlike the rasters you interpolated earlier, the Natural

Neighbors surface doesn't come with default elevation color scheme. The symbology is also stretched instead of classified.

Step 8
Symbolize the Natural Neighbors Surface

In the Table of Contents, double-click *nngrid* to bring up its layer properties.

Click the Symbology tab. In the Show box, click Classified.

In the Classification panel, change the number of classes to 10.

Click the Classify button and change the classification method to Equal Interval. Click OK to close the Classification dialog.

Right-click the color ramp, and uncheck Graphic View. Scroll down and select the Surface color ramp.

On the Base Heights tab, click *Obtain heights for layer from surface,* and choose the *NNGRID* raster from your *Chapter04\ MyData* folder.

On the Rendering tab, check the box to shade areal features.

Click OK to close the Layer Properties dialog.

Step 9
Change View Settings and Scene Properties

From the View menu, choose View Settings. Change the projection back to Perspective.

Close the dialog.

In the Table of Contents, double-click Scene Layers to bring up the scene properties.

On the General tab, change the Vertical Exaggeration to 5.

Click OK.

Now the Natural Neighbors raster looks more recognizable, its estimations falling somewhere between the over-localization of IDW Power 2 and the over-generalization of IDW Power 0.4. Zoom in to take a look at the new surface.

Step 10
Close ArcScene and Save the Document

Feel free to experiment with the various interpolators. When you're finished, save the ArcScene document under a name of your choice in your *GTK3D\Chapter04\MyData* folder.

EXERCISE 3: CALCULATE HILLSHADE AND ASPECT

In this exercise, you'll create a hillshade and an aspect raster from a DEM of a portion of San Luis Obispo County.

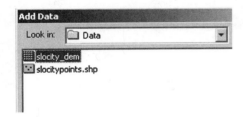

Step 1
Start ArcScene and Add Data

Open ArcScene and click the Add Data button.

Navigate to your *GTK3D\Chapter04\Data* folder.

Add *slocity_dem* to the scene.

Slocity_dem is a DEM covering a larger area of San Luis Obispo County than you worked with in the last exercise.

Step 2
Look at slocity_dem's Attributes

In the Table of Contents, right-click *slocity_dem* and choose Open Attribute Table.

Slocity_dem has 805 records, with elevation values ranging from 19 to 826. It was converted to an integer raster from a DEM with floating-point values.

Close the attribute table.

Step 3
Set the Analysis Environment

From the 3D Analyst menu, choose Options.

On the General tab, set the Working directory to *GTK3D\Chapter04\ MyData*.

Set the Analysis mask to None.

On the Extent tab, set the Analysis extent to *Same as Layer slocity_dem*.

On the Cell Size tab, set the Analysis cell size to *Same as Layer slocity_dem*.

Click OK.

Step 4
Calculate Hillshade

On the 3D Analyst menu, choose Surface Analysis, and then Hillshade.

The input surface should be set to *slocity_dem*.

You may remember that azimuth is the compass direction of the sun, and that altitude is the angle of the sun above the horizon. Here, the default azimuth value of 315 degrees places the sun in the northwest, and an altitude of 45 puts it halfway between the horizon and the zenith.

As discussed earlier, a hillshade calculates the illumination of each cell on a surface from a hypothetical light source. Cells can be given values from 0 (black) to 255 (white). The Model Shadows option intensifies shadowing by assigning a value of 0 to all cells that are in the shadow of another cell, instead of giving them some value between 1 and 255. Shadow modeling is used to identify cells that will be in the shadow of another cell at a particular time of day.

The Z factor is a multiplier used when an input layer's *x,y* units are in a different measurement system than its *z* units. For example, if the *x,y* units are meters and the *z* units are feet, a Z factor of 3.28 converts the *z* units to meters (1 foot x 3.28 = 1 meter).

Click the Browse button next to Output Raster and navigate to your *GTK3D\Chapter04\MyData* folder. Name the hillshade *hillshd1* and click Save.

Leave the other parameters at their default values. The Hillshade dialog should look as shown at left.

Click OK.

The new hillshade raster is added to the scene.

Notice that even though no base heights have been set for either the DEM or the hillshade raster, it looks three-dimensional. Hillshading is often used to add the effect of terrain relief to 2D data. By placing a semitransparent elevation raster on top of a hillshade, you can create an even more realistic image of the landscape.

Step 5
Set Symbology and Transparency for slocity_dem

In the Table of Contents, double-click *slocity_dem* to open its layer properties.

On the Symbology tab, right-click the color ramp and uncheck Graphic View. Scroll down and select the ramp called Yellow to Green to Dark Blue.

On the Display tab, type *40* in the Transparent box.

Click OK to close *slocity_dem*'s layer properties.

Step 6
Set Drawing Priority for hillshd1

In the Table of Contents, double-click *hillshd1* to bring up its layer properties.

Click the Rendering tab. In the Effects panel, choose a drawing priority of 2.

Click OK.

The partially transparent DEM draws over the hillshade, giving an impression of topographic relief.

Turn *hillshd1* off and back on in the Table of Contents to see the effect of the combined rasters. This way of showing relief is commonly used in orthographic displays—for example, in ArcMap, where 3D viewing isn't available.

Step 7
Calculate Aspect

As you may remember, aspect identifies the steepest downslope direction from each cell to its neighbors. It can be thought of as slope direction or the compass direction a hill faces.

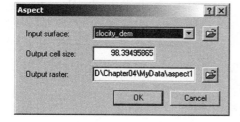

From the 3D Analyst menu, choose Surface Analysis, and then Aspect.

Make sure the input surface is set to *slocity_dem*. (An aspect raster calculated from *hillshd1* would look pretty odd, although feel free to try it out at the end of the exercise.)

Name the output raster *aspect1*, and save it in your *GTK3D\Chapter04\MyData* folder.

Click OK.

The new aspect raster is added to the scene.

The aspect raster is automatically symbolized with nine colors for the eight compass directions plus gray for flat areas. Red symbolizes North, yellow is East, medium blue is Southwest, and so on.

Step 8
Set Base Heights for All Three Layers

In the Table of Contents, double-click *aspect1* to bring up its layer properties.

On the Base Heights tab, click *Obtain heights for layer from surface* and select *slocity_dem*.

Click OK.

Do the same for *slocity_dem* and *hillshd1*, setting the base heights in each case to *slocity_dem*.

Step 9
Set Vertical Exaggeration

In the Table of Contents, double-click Scene Layers.

Set the Vertical Exaggeration to 3, and then click OK.

Zoom in and navigate around the surfaces, turning them on and off to see the different effects that hillshade, hillshade plus elevation symbology, and aspect create.

Step 10
Save the Document

When you're finished experimenting, give the scene a name of your choice and save it in your *GTK3D\Chapter04\MyData* folder. Then close ArcScene.

EXERCISE 4: CALCULATE SLOPE

Knowing the slope of an area is important for many applications, from modeling stream runoff to predicting mudslides to delineating flood plains. In this exercise, you'll create a slope raster of terrain around Santa Margarita Lake in California. You'll also reclassify a raster, and then use it as an analysis mask to keep the lake from being included in the slope model.

Step 1
Start ArcScene and Add Data

Open ArcScene and click the Add Data button.

Navigate to your *GTK3D\Chapter04\Data* folder.

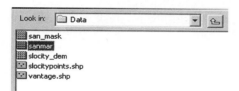

Add the *sanmar* raster to the scene.

Sanmar is a DEM of the area surrounding Santa Margarita Lake in San Luis Obispo County.

Step 2
Set the Analysis Environment

From the 3D Analyst menu, choose Options.

On the General tab, set the Working directory to *GTK3D\Chapter04\MyData*.

The Analysis mask should be set to None.

On the Extent tab, the Analysis extent should default to Intersection of Inputs. On the Cell Size tab, the Cell size should be set to Maximum of Inputs.

Click OK.

Step 3
Calculate Slope

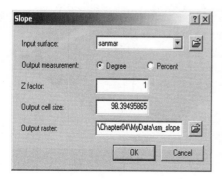

From the 3D Analyst menu, choose Surface Analysis, and then Slope.

Click the Browse button next to the Output raster path name. Navigate to your *GTK3D\Chapter04\MyData* folder and name the new raster *sm_slope*.

Leave the other parameters as they are. You could calculate a raster that showed either percent or degree of slope, but in this case we'll stick with the latter.

Click OK.

A new slope raster is added to the scene.

In the Table of Contents, turn off *sanmar*.

Zoom in and take a look at the slope raster. Each cell in the terrain is symbolized by its degree of slope, with steeper slopes in dark red and flat regions in dark green.

When you're done, collapse the legend for *sm_slope*.

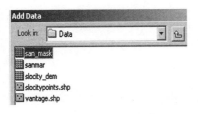

Step 4
Add a Mask Layer

Zoom to the full extent of the scene.

Click the Add Data button, navigate to your *GTK3D\Chapter04\Data* folder, and add *san_mask* to the scene.

Turn off *sm_slope* in the Table of Contents.

Step 5
Look at san_mask's Attributes

Right-click *san_mask* in the Table of Contents and choose Open Attribute Table.

San_mask is a raster covering the same area as *sanmar*, but with only two cell values: 1 and *NoData*. The *NoData* values are not shown in the attribute table; all you can see is that there are 3,234 cells with a value of 1. These cells correspond to Santa Margarita Lake.

Close the attribute table.

Step 6
Symbolize san_mask

3D Analyst uses a transparent "color" to symbolize *NoData* values. You can give them a color, however, to get a better idea of the geographical extent of *san_mask*.

In the Table of Contents, right-click the symbol under *san_mask* to bring up the Color palette.

Choose Dark Amethyst.

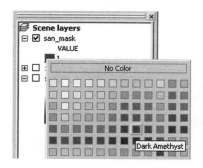

Now double-click *san_mask* to bring up its layer properties.

Click the Symbology tab.

Click the drop-down arrow next to *Display NoData as* and choose Chrysophase.

Click OK to close the Layer Properties dialog.

Now you can see *san_mask*'s *NoData* values in the scene.

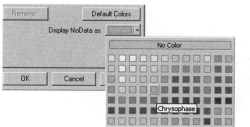

An analysis mask is a layer used to keep certain areas from being processed during the interpolation or reclassification of a new raster. It can be either a vector layer (point, line, or polygon) or a raster. Wherever the raster is covered by the mask, it is processed. Areas outside the mask, or areas inside the mask valued *NoData*, are not processed.

What you'd like to do is create a new slope raster, but you don't want slope values calculated into the lakes.

In our case, *san_mask* has values of 1 wherever there is water, and values of *NoData* for the rest of the surface. If you used the mask as it is, the new slope raster would only be processed where the lake is, which is not what you want. So, you're going to create a new mask layer, reclassified so that the land has a value of 1 and the water body has a value of *NoData*. This way, your new slope raster will be calculated for the land only, leaving a blank area of *NoData* to demarcate Santa Margarita Lake.

Step 7
Check the Analysis Environment

When you're reclassifying or interpolating a surface, it's always good to check the analysis environment. 3D Analyst holds onto some of the previously set parameters, such as cell size, even if they were last set in an entirely different scene. If you create a surface and it comes out with the wrong extent or cell size, or gets saved in the wrong location, it probably has to do with the settings in the analysis environment.

From the 3D Analyst menu, choose Options.

On the General tab, make sure the Working directory is set to *GTK3D\Chapter04\MyData*.

The Analysis mask should be set to None. (You're making the mask in this step, not using it yet.)

On the Extent tab, set the Analysis extent to *Same as Layer san_mask*.

On the Cell Size tab, also set the Analysis cell size to *Same as Layer san_mask*.

Click OK to close the Options dialog.

Step 8
Reclassify san_mask to Create a New Mask Layer

From the 3D Analyst menu, select Reclassify.

Set the Input raster to *san_mask*.

In the New Values panel, type *NoData* to replace the old value of 1, and type *1* to replace the old value of *NoData*.

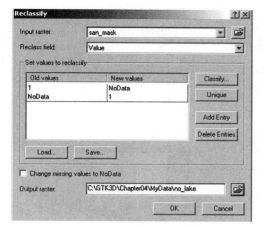

Name the Output raster *no_lake* and save it in your *GTK3D\Chapter04\MyData* folder.

Click OK.

The new raster is added to the scene.

In the Table of Contents, turn off *san_mask* so that you can see *no_lake* undisguised.

No_lake has the value 1 wherever there is land, and the value *NoData* wherever there is water. By default, the *NoData* values are transparent. This is the mask you will use to calculate slope in the next step.

It's worth noting the difference between reclassifying the cell values in a raster and merely changing its classification scheme. When you use the Symbology tab to change the number of classes displayed in a raster (five classes to nine, for instance) or to change its classification method (say, Equal Interval to Jenks) you are not changing the raster cell values. All you change is the scheme by which the values are symbolized.

This process is commonly called classification, which can be confusing because "reclassification" is the process of changing actual cell values after their initial interpolation. When you use the 3D Analyst menu to create a slope, aspect, hillshade, or other reclassification, you are making a new raster with new cell values. When you gave the *NoData* cells a value of 1 and the 1 cells a value of *NoData*, you altered the cell values: you changed the data. Changing the classification *scheme* of a raster doesn't change the data, it just changes the way it's represented in the scene. Generally, if you change the data, you are making a new raster.

Step 9
Change the Analysis Environment

From the 3D Analyst menu, choose Options.

On the General tab, make sure the Working directory is set to *GTK3D\Chapter04\MyData*.

Set the Analysis mask to *no_lake*.

On the Extent tab, set the Analysis extent to *Same as Layer sanmar*.

On the Cell Size tab, also set the Analysis cell size to *Same as Layer sanmar*.

Click OK to close the Options dialog.

Step 10
Calculate Slope with the no_lake Mask

Turn off all layers in the Table of Contents.

From the 3D Analyst menu, choose Surface Analysis, and then Slope.

Set the Input surface to *sanmar*.

Name the Output raster *sm_slope2* and save it in your *GTK3D\Chapter04\MyData* folder.

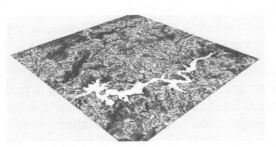

Click OK.

The new slope raster is added to the scene.

Slope was calculated wherever *no_lake* had cell values. The *NoData* values were also carried over to the new raster.

Step 11
Set Layer Properties for sm_slope2

In the Table of Contents, double-click *sm_slope2* to bring up its layer properties.

On the Base Heights tab, click *Obtain heights for layer from surface* and choose *sanmar*.

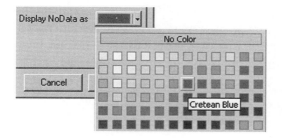

On the Rendering tab, check the box to shade areal features.

On the Symbology tab, click the drop-down arrow next to *Display NoData as* and this time choose Cretean Blue.

Click OK to close the dialog.

Step 12
Set Scene Properties

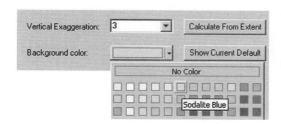

In the Table of Contents, double-click Scene Layers.

On the General tab, set the Vertical Exaggeration to 3.

Set the Background color to Sodalite Blue.

Click OK to close the dialog.

Step 13
Navigate the Scene

Zoom in and take a look at the new slope layer. When you're finished, give the Arc-Scene document a name of your choice and save it in your *GTK3D\Chapter04\MyData* folder. Then close ArcScene.

EXERCISE 5: CALCULATE VIEWSHED

You learned in our earlier discussion that a viewshed raster identifies the areas of an elevation surface that can be seen from one or more observation points, and that each cell in the viewshed stores the number of observers who can see it. If you have only one observation point, each cell that can be seen from that point gets a value of 1. Cells that can't be seen from that point get a value of 0.

If more than one observer point is used, each visible cell in the viewshed stores the number of observers who can see it. Viewshed analysis assumes that the observer's field of view has no horizontal or vertical constraints, nor any distance limits.

In this exercise, you'll create viewsheds of several observation points around the city of San Luis Obispo.

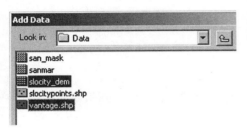

Step 1
Start ArcScene and Add Data

Open ArcScene and click the Add Data button.

Navigate to your *GTK3D\Chapter04\Data* folder.

Add *slocity_dem* and *vantage.shp* to the scene.

Step 2
Symbolize slocity_dem

In the Table of Contents, double-click *slocity_dem* to bring up its layer properties.

On the Base Heights tab, click *Obtain Heights for Layer from Surface* and choose *slocity_dem*.

On the Symbology tab, right-click the color ramp and uncheck Graphic View. Scroll down and choose Brown to Blue-Green Diverging, Bright.

On the Rendering tab, check the box to shade areal features.

Click OK to close the dialog.

Step 3
Set Vertical Exaggeration and Background Color

In the Table of Contents, double-click Scene Layers to bring up Scene Properties.

On the General tab, set the Vertical Exaggeration to 3, and set the Background Color to a light blue of your choice.

Click OK to close the dialog.

Step 4
Symbolize the Vantage Points

In the Table of Contents, left-click the dot symbol for the *vantage* shapefile.

In the Options panel of the Symbol Selector, change the symbol size to 7 and change the color to a bright red of your choice.

Click OK to close the dialog.

Step 5
Examine the Point Shapefile's Attribute Table

In the Table of Contents, right-click *vantage* and choose Open Attribute Table.

Vantage.shp has five records. Their heights are listed in the Elevation field. The *Point_ID* field lists the observation points in order of their elevation: 1 is the highest, 5 is the lowest. Their *x* and *y* coordinates are displayed as well.

Close the attribute table.

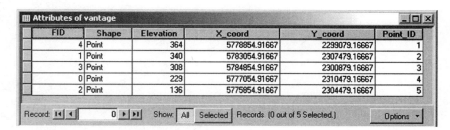

FID	Shape	Elevation	X_coord	Y_coord	Point_ID
4	Point	364	5778854.91667	2299079.16667	1
1	Point	340	5783054.91667	2307479.16667	2
3	Point	308	5784854.91667	2300879.16667	3
0	Point	229	5777054.91667	2310479.16667	4
2	Point	136	5775854.91667	2304479.16667	5

Step 6
Set Base Heights for the Vantage Points Layer

Double-click *vantage* to bring up its layer properties.

On the Base Heights tab, click *Use a constant value or expression to set heights for layer*. Click the Calculator button to bring up the Expression Builder.

You could use *slocity_dem* to set the base heights for the vantage points, but since the points have their own elevation values listed conveniently in their attribute table you'll use those instead.

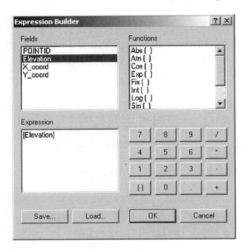

In the Expression Builder, select Elevation in the Fields box. It should appear in the Expression box, with brackets around it.

You're telling 3D Analyst to use the five values in the attribute table's Elevation field as the base heights for *vantage.shp*.

Click OK in the Expression Builder, and then click OK again to close the Layer Properties dialog.

The vantage points are symbolized by their elevation values in the scene, nestled in the hillsides of *slocity_dem*.

Step 7
Set the Working Directory

From the 3D Analyst menu, choose Options.

On the General tab, set the Working directory to your *GTK3D\ Chapter04\MyData* folder.

Make sure the Analysis mask is set to None.

On the Extent tab, set the Analysis extent to *Same as Layer slocity_dem.*

On the Cell Size tab, set the analysis cell size to *slocity_dem* as well.

Click OK to close the dialog.

Step 8
Select the Highest Observation Point

In the Table of Contents, right-click *vantage* and choose Open Attribute Table.

Select the record for *Point_ID 1* by clicking on its gray square at the left edge of the table.

FID	Shape	Elevation	X_coord	Y_coord	Point_ID
0	Point	229	5777054.91667	2310479.16667	4
1	Point	340	5783054.91667	2307479.16667	2
2	Point	136	5775854.91667	2304479.16667	5
3	Point	308	5784854.91667	2300879.16667	3
4	Point	364	5778854.91667	2299079.16667	1

Point_ID 1 has the highest elevation. You're going to create a viewshed raster of all cells in *slocity_dem* that can be seen from that vantage point. Notice that the location has also been highlighted in the scene.

Leave the record selected, and minimize the attribute table.

Step 9
Calculate Viewshed

From the 3D Analyst menu, choose Surface Analysis, and then Viewshed.

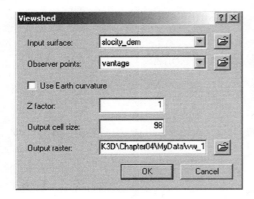

The Input surface is *slocity_dem*, and the Observer points layer is vantage. Only the selected point will be used in the calculation.

Name the Output raster *vw_1* and save it in your *GTK3D\Chapter04\MyData* folder.

Click OK.

The viewshed from the observation point is calculated and added to the scene.

Step 10
Set Layer Properties for vw_1

In the Table of Contents, turn off *slocity_dem* so that you can see the new viewshed raster.

Double-click *vw_1* to bring up its layer properties.

On the Base Heights tab, click *Obtain heights for layer from surface*, and then choose *slocity_dem*.

On the Rendering tab, check the box to shade areal features.

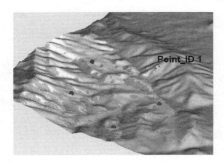

On the Display tab, change the transparency to 0%.

Click OK to close the Layer Properties dialog.

Green cells indicate areas that can be seen from *Point_ID 1*; pink cells indicate areas that can't. Navigate around the surface to get an idea of *Point_ID 1*'s perspective.

Step 11
Examine vw_1's Attribute Table

Right-click *vw_1* and choose Open Attribute Table.

ObjectID	Value	Count
0	0	125583
1	1	17844

The cells that have a value of 1 are visible from the observation point; the cells with a value of 0 are not. The Count field tells you how many there are of each.

Close the attribute table.

In a slope or aspect calculation, small errors in cell values tend to average out in the final result, so the quality of the elevation surface doesn't matter as much. The directional compass used to measure aspect and the basic rules of geometry used to define slope are always the same. Viewshed computation, however, tends to magnify any errors in the elevation surface. The observation points from which a viewshed is calculated are always variable, so errors are more easily made and tend to compound one another. Incorrect x,y coordinates of the observer points, incorrect z values on the elevation surface, and incorrect viewshed parameters can have a snowballing effect on cell calculations.

Moving an observation point a few meters can result in a very different viewshed calculation. Also, if a cell directly in front of an observer point is incorrectly given a value of 0, other cells that would have been visible will be ruled out by the value of the first incorrect cell. Finally, snapping or not snapping the output extent to the input raster can also make a considerable difference in the results.

Step 12
Clear the Selected Feature from vantage.shp

In the Table of Contents, right-click *vantage*, click Selection, and then choose Clear Selected Features.

Resize the attribute table for the vantage points shapefile. (If you closed it earlier, reopen it.)

You're going to make a new viewshed raster using all five observation points, so make sure that none of the five records are selected.

Close the *vantage* attribute table.

Step 13
Calculate Viewshed for the Observer Points

From the 3D Analyst menu, choose Surface Analysis, and then Viewshed.

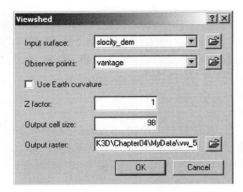

Make sure the Input surface is *slocity_dem*.

Name the Output raster *vw_5* (for five observation points) and save it in your *GTK3D\ Chapter04\MyData* folder.

Click OK.

The process takes a few moments. When it is finished, *vw_5* is added to the scene.

Step 14
Set Layer Properties for vw_5

Turn off *vw_1* in the Table of Contents.

Double-click *vw_5* to bring up its layer properties.

On the Base Heights tab, check *Obtain heights for layer from surface* and select *slocity_dem*.

On the Rendering tab, check the box to shade areal features.

On the Display tab, set the transparency to 0.

Click OK to close the dialog.

This time, the green areas can be seen from at least one of the five lookout points. Pink areas can't be seen from any point.

Step 15
Symbolize the Viewshed Layer

Double-click *vw_5* to bring up its layer properties, and then click the Symbology tab.

In the Show box, click Unique Values.

Right-click the color bar in the Color Scheme drop-down box and uncheck Graphic View.

Choose Enamel from the Color Scheme menu.

Below the Symbol column, double-click the color symbol next to 0, to bring up the Color palette.

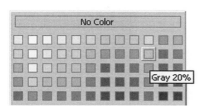

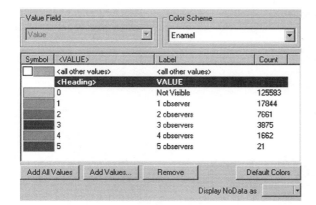

Choose Gray 20%.

In the Label column, click the label *0*. In the highlighted text box that results, type *Not Visible*.

Press Enter.

Replace the label *1* with *1 observer*, the label *2* with *2 observers*, and so forth. Press Enter when you're done with each.

Click OK to close the dialog.

Now the new viewshed raster is symbolized by the number of observers who can see each portion of it. Note that it is not symbolized by which areas can be seen by each observer. To do that, you would have to make five viewsheds, each of the areas seen by one observer, and overlay them.

Step 16
Save the Document

Navigate the scene and take a gander at the perspectives from various observation points. If you like, use the Set Target and Set Observer tools to test the view of the raster from the observation points themselves.

When you're finished, give the scene a name of your choice and save it in your *GTK3D\Chapter04\MyData* folder. Then close Arc-Scene.

CHAPTER 5

TIN SURFACE MODELS

In Chapter 1 you learned that a TIN (triangulated irregular network) is a vector data structure that represents a surface by dividing geographic space into contiguous, nonoverlapping triangles, and that each triangle node stores an *x*, *y*, and *z* value.

TIN INTERPOLATION

Like a raster, a TIN is created from sample points, and their elevation values are interpolated to form a surface. A TIN is a little less complex, however, than a raster. A TIN is almost always created from elevation samples, whereas a raster can be created from samples of any phenomenon that varies continually. Further, when you create a TIN you have only one interpolation method to choose from; specifically, Delaunay triangulation.

The Delaunay method generates vector triangles from sample points so that the points become the nodes of triangles. The triangles are arranged so that a circle drawn around any triangle can contain no other nodes. This means that collectively the triangles are constrained to be as equiangular as possible. Like many geometric definitions, this is best explained with a picture:

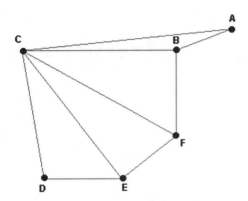 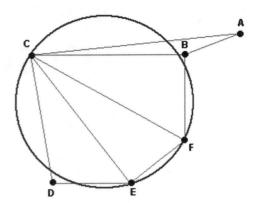

A circle drawn through the nodes of triangle CEF neither contains nor passes through any other points. This is the case for each triangle in the TIN. This keeps the triangles small and fat, which is good because long, narrow triangles are difficult for computers to process. Another feature of the Delaunay rule is that the TIN will come out the same no matter what order the points are processed in.

The sample points used to create a TIN are called "mass points" because they define the bulk, or mass, of the TIN. The samples can come from point features, or from the vertices of line or polygon features. Their elevation values can come from the feature layer's attribute table or, if it's a 3D layer, from z values in the Shape field. The sample points form the triangle nodes, and the triangulation consists of connecting the nodes by lines. Once the TIN is built, the elevation of any location on its surface can be estimated using the x, y, and z values of the bounding triangle's vertices. Each triangle in the TIN is a plane and therefore has one slope value and one aspect value, but the elevation on any part of a triangle face can be calculated from the surrounding nodes.

When you identify any location on the face of a TIN, the x, y, and z values of the surrounding triangle nodes are used to interpolate the elevation at that point. The nodes are also used to calculate the slope and aspect of each triangle face.

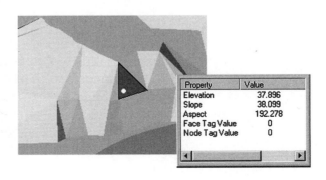

Like a raster, a TIN is only as good as the initial sample points taken. Real terrain often has shapes that aren't well represented by mass point triangulation. Ridges, cliffs, lakes, and gullies are a few examples. To model these areas, you can include line and polygon features when you interpolate the TIN.

BREAKLINES, REPLACE POLYGONS, CLIP POLYGONS, ERASE POLYGONS, AND FILL POLYGONS

Regular mass point triangulation (below left) doesn't adequately model the river. Breaklines (below right) are used to give it definition.

There are several ways polylines and polygons can add definition to a TIN. Breaklines are lines that tell the interpolation engine that there is a distinct change in slope on either side of the line, and that no triangle should cross it. Since locations along opposite sides of the breakline belong to different triangles, they have different slope values (as long as the surface isn't flat).

Breaklines are used to represent surface formations such as ridges, streams, dams, shorelines, and building footprints. They may or may not contain elevation values. If they do, the line vertices are used as mass points.

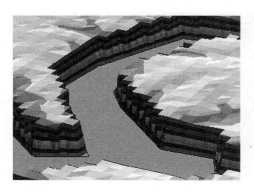

A replace polygon creates a flat area on a TIN surface. It stores one elevation value as an attribute. When you add the polygon, the TIN is retriangulated so that the surface area covered by the polygon has the new value.

A polygon of a reservoir is used to make the surface flat and give the banks appropriate definition.

A clip polygon trims the boundary for which elevation, slope, and aspect are calculated. Clipping doesn't actually change the extent of the triangulated area, however; it just keeps TIN faces from being calculated. It does affect analysis, because the statistics for area, volume, elevation, slope, and aspect are calculated for fewer faces. Since their boundaries become triangle edges, clip polygons are often used to give a TIN a more regular shape. They may or may not contain elevation values.

 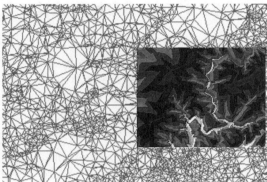

The polygon within the left-hand image is used to clip out a zone of interpolation from the TIN. The entire TIN is still triangulated, but only the clipped area has calculations for elevation, slope, and aspect. (Normally, the additional triangles aren't symbolized, but showing them here explains why, in a scene, the TIN extent will be larger than the visible clipped area.)

Erase polygons work just like clip polygons, but in reverse. They cut a hole in the TIN instead of trimming it.

Here, the same polygon has been used to erase the area it overlays. Again, the entire TIN is still triangulated, but elevation, slope, and aspect are calculated only for the area outside the polygon.

Fill polygons tag areas of the TIN so that the TIN can be symbolized by attributes other than its main theme of elevation, slope, or aspect. The attributes must be expressed by integer values. Fill polygons can symbolize surface features such as land cover, land use, flood zones, and endangered species habitats. They cause the TIN to be retriangulated, but they don't change its elevation, slope, or aspect values.

The term *fill polygon* seems a lot like the term *replace polygon*, but they're quite different. A replace polygon replaces an area of a TIN with a new elevation value, such as a flat lake bottom. A fill polygon, which probably should have been called an "attribute polygon," adds a tag value of some attribute to the existing TIN without changing its shape.

Vegetation polygons have been added as fill polygons to the TIN. Elevation values aren't affected; the TIN retains its shape, and areas of plant life are clearly marked.

You can also assign attributes to triangle faces by specifying a tag value field from the attribute table of replace or clip polygons, and attributes can be assigned to triangle nodes by specifying a tag value from the attribute table when you add mass points. In all cases, the attribute values have to be integers. You can't use tag values with breaklines or with erase polygons.

The chart below summarizes the surface feature types that can be added to a TIN, and which ones require elevation values. It comes in handy when you're trying to remember the difference between a replace, an erase, and a fill polygon.

Surface feature type	Purpose	Input layer	Elevation information	Tag Values
Mass points	Primary source of elevation values	Point, line, or polygon	Required	Optional
Breaklines	Enforce changes in slope	Line or polygon	Optional	Unavailable
Clip polygons	Define zone of interpolation	Polygon	Optional	Optional
Erase polygons	Define zone of interpolation	Polygon	Optional	Unavailable
Replace polygons	Replace with a constant elevation	Polygon	Required	Optional
Fill polygons	Assign attributes to triangles	Polygon	Optional	Purpose

TIN Symbology

In Chapter 4 you learned to classify raster surfaces by elevation, slope, and aspect, and you saw that 3D Analyst provides specific color schemes to symbolize those themes. TINs are very similar, but simpler. When you reclassify a raster surface, you have to choose whether to create a new slope, elevation, or aspect raster, and 3D Analyst symbolizes it accordingly. A TIN already contains all of this information, so all you have to do is choose which of the themes you want to symbolize.

The same TIN symbolized from top to bottom by elevation, slope, and aspect.

In the following exercises you'll create a TIN from contour lines, add attribute values to it, and change its classification scheme.

EXERCISE 1: CREATE A TIN FROM VECTOR FEATURES

In this exercise you'll use a shapefile of contour lines to create a TIN of a local knoll on the outskirts of San Luis Obispo.

Step 1
Start ArcScene and Add Data

Open ArcScene and click the Add Data button.

Navigate to your *GTK3D\Chapter05\Data* folder.

Add *contour_10ft.shp* to the scene.

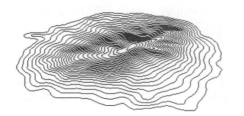

Contour_10ft is a shapefile of contour lines representing Cerro San Luis, also called San Luis Mountain (and, occasionally, Cerro San Luis Obispo). It's one of a chain of small volcanic peaks that extend from Morro Bay to San Luis Obispo.

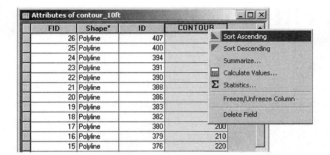

Step 2
Examine the Shapefile's Attribute Table

In the Table of Contents, right-click *contour_10ft* and choose Open Attribute Table.

Right-click the CONTOUR field and choose Sort Ascending.

Examine the table. The contour values range from 110 to 370 feet above sea level.

Close the table.

Step 3
Set Base Heights for contour_10ft

In previous exercises, you used raster and TIN surfaces to set the base heights for vector layers. Since contour lines have their own elevation values, you can use the values in the CONTOUR field to set the shapefile's base heights.

In the Table of Contents, right-click *contour_10ft* and choose Properties.

On the Base Heights tab, click "Use a constant value or expression..." and click the Calculator button.

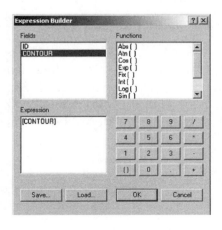

In the Expression Builder, click CONTOUR in the Fields box. It should appear, with brackets around it, in the Expression box.

Click OK.

In the Base Heights tab, the Height panel should now be set to use the values in the CONTOUR field for base heights.

Click OK again to close the Layer Properties dialog.

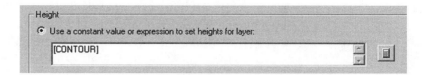

Step 4
Set Vertical Exaggeration

If you zoom in and navigate around the contour lines, you can see that their base heights are set. Still, since the hill is about a mile in diameter an overall elevation of 260 feet isn't very noticeable, so you'll give it a little lift.

In the Table of Contents, double-click Scene Layers to bring up the Scene Properties dialog.

In the General tab, change the Vertical Exaggeration to 3, and set the Background to a color of your choice.

Click OK.

Step 5
Create a TIN

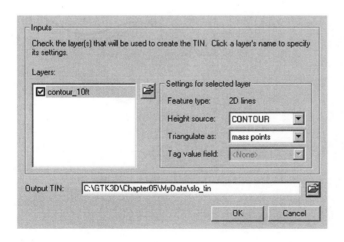

From the 3D Analyst menu, choose Create/Modify TIN, and then choose Create TIN From Features.

In the dialog under Layers, check *contour_10ft*.

Under Settings, set the height source to CONTOUR, and triangulate it as mass points. This tells 3D Analyst to treat all vertices in the contour lines as a mass of *x,y,z* locations.

Leave the Tag Value field set to None. Name the output TIN *slo_tin* and save it in your *GTK3D\Chapter05\MyData* folder.

Click OK.

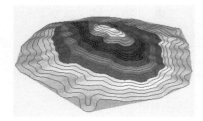

The TIN is added to the scene, displayed with a default elevation symbology and with its base heights already set. Since a TIN almost always represents an elevation surface, 3D Analyst assumes that this is what you want. (A raster, as you may remember, can represent many types of phenomena besides elevation, so 3D Analyst leaves setting raster base heights up to you.)

Step 6
Examine the TIN

Zoom in and navigate the surface of the TIN. Turn the contour lines on and off in the Table of Contents to get an idea of how the TIN was made from the elevation values at the vertices of each line. Also, notice that the boundary of the TIN isn't clipped to the outside contour line but is instead triangulated into a more general shape around the TIN.

Step 7
Symbolize the TIN

In the Table of Contents, double-click *slo_tin* to bring up its layer properties, and then click the Symbology tab.

Right-click the color ramp in the drop-down list, and uncheck Graphic View.

Scroll up and choose the Blue-Green Bright color ramp.

Click OK.

Step 8
Add the Hill's Boundary Polygon

Click the Add Data button.

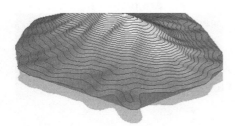

Navigate to your *GTK3D\Chapter05\Data* folder, and add *slo_hill_bnd.shp* to the scene.

Slo_hill_bnd is a polygon shapefile made by unioning a square polygon of the area around the hill with a narrow buffer made from the outermost contour line. After some selection, exporting of data, dissolving, and a bit of cursing, the final shapefile is a polygon bounding the hill's outside contour, buffered at about 30 feet.

Step 9
Clip the TIN to the Contour Boundary

From the 3D Analyst menu, choose Create/Modify TIN, and then Add Features to TIN.

The input TIN is set to *slo_tin*.

In the Layers panel, check the box next to *slo_hill_bnd*.

Set the Height Source to None, and triangulate the layer as a hard clip.

Leave the tag value set to None, and make sure the changes will be made to the existing input TIN.

Click OK. After a few moments, the TIN is clipped to the shape of the polygon.

Turn off *contour_10ft* and *slo_hill_bnd* in the Table of Contents.

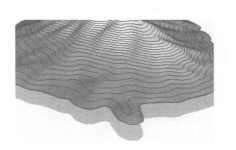

When you first created the TIN, you could have triangulated the sample points and clipped the TIN in one step instead of two. The Create TIN from Features dialog lets you set the parameters for each layer involved, and then does the operation all at once. To keep things from getting too confusing in this first illustration, though, it made more sense to perform the operations separately.

Step 10
Navigate the Hill and Use the ID Tool

Use the Zoom In/Out and Navigate tools to take a look at the TIN you've created. Click the ID button and click anywhere on the TIN.

The ID tool tells you, for any location that you click on, the TIN's elevation, slope, and aspect. A TIN face has only one slope value and aspect value, but the elevation varies along the face, and 3D Analyst interpolates this value using the face's three nodes. If you zoom in far enough to click on several different locations within a single face, you'll get the same slope and aspect value each time, but different elevation values.

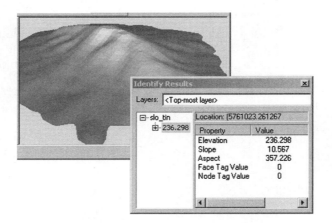

Step 11
Close ArcScene

When you're finished using the ID tool, give the scene a name of your choice and save it in your *GTK3D\Chapter05\MyData* folder.

EXERCISE 2: ADD POLYGON ATTRIBUTE VALUES TO A TIN

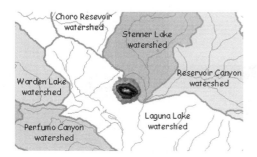

The peak of Cerro San Luis divides the Stenner Lake watershed from the Laguna Lake watershed in San Luis Obispo County. In this exercise you'll symbolize the TIN of Cerro San Luis with polygons of the watershed boundaries.

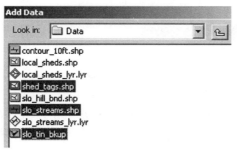

Step 1
Start ArcScene and Add Data

Open ArcScene and click the Add Data button.

Navigate to your *GTK3D\Chapter05\Data* folder, and add *shed_tags.shp*, *slo_streams.shp*, and *slo_tin_bkup* to the scene.

Step 2
Examine the Data

In the Table of Contents, right-click *slo_tin_bkup* and choose Zoom to Layer. This is a copy of the TIN you made in the last exercise.

Shed_tags lies underneath the TIN. It's a polygon layer of the Stenner and Laguna Lake watersheds, clipped to the extent of Cerro San Luis.

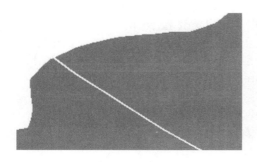

Turn off *slo_tin_bkup* in the Table of Contents so that you can see the polygons. Zoom in to the split between them. It's a very small distance, only about 2 meters. The split indicates the ridgeline that separates the two watersheds.

Open the *shed_tags* attribute table. The table lists the acreage and names of the watersheds. It also includes a Tag field with a code value for each watershed, which you'll use when you incorporate the polygons into the TIN. (Tag values have to be integers, so you can't just tag the TIN with the names of the watersheds.)

Close the attribute table.

FID	Shape*	NAME	ACRES	TAG
0	Polygon	Stenner Lake	7259	100
1	Polygon	Laguna Lake	18172	200

Record: ◄◄ ◄ 0 ► ►► Show: All Selected Records (0 out of 2 Selected.)

Step 3
Set Base Heights for the Polygons

Turn *slo_tin_bkup* on again.

Zoom to the extent of *shed_tags*, and open its layer properties.

On the Base Heights tab, click *Obtain Heights for layer from surface* and choose *slo_tin_bkup* as the source.

In the Offset panel, type in a value of *150* so that the polygons won't compete with the TIN for the same 3D space.

Click OK.

The watershed polygons are displayed with the TIN's base heights and an offset of 150 feet, but you can still see the TIN peeking through. You could turn the TIN off, or give it a lower drawing priority, but the polygons still would not have the TIN's quality of detail. That's because base heights for vector features can be set for their vertices only, which works fine for points or lines but not as well for the vertex-free area inside a polygon.

Step 4
Add the Polygon Tag Attributes to the TIN

Zoom back to the extent of *slo_tin_bkup* and *shed_tags*.

From the 3D Analyst menu, choose Create/Modify TIN, and then Add Features to TIN.

In the Layers panel, check the box next to *shed_tags*.

Set the height source to None, triangulate it as a soft value fill, and set the tag value to TAG.

Save it as a new output TIN in your *GTK3D\Chapter_05\MyData* folder, and name it *slo_tin_shed*.

Click OK.

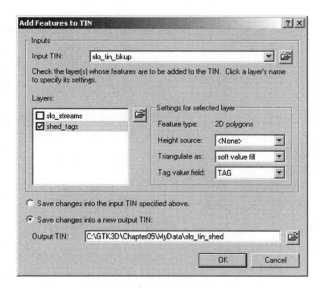

After a while, the new TIN is added to the scene. It looks just like *slo_tin_bkup* but with a line running down the middle.

Step 5
Symbolize the TIN by the Watershed Polygons

Remove *shed_tags* and *slo_tin_bkup* from the Table of Contents.

Right-click *slo_tin_shed* and bring up its layer properties.

On the Symbology tab, uncheck Elevation and Edge Types in the Show box.

Click the Add button and choose *Face tag value grouped with unique symbol*.

Click Add, and then click Dismiss.

The Face tag values are added to the Show box, and should be highlighted.

Next, uncheck the box next to *< all other values >* in the main Symbol panel.

Click Add All Values.

Three attribute symbols are added: 100 for the Stenner Lake watershed, 200 for the Laguna Lake watershed, and 0 for the rest (essentially a *NoData* value).

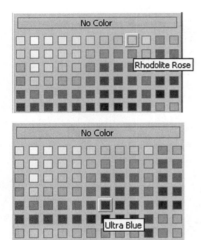

Double-click the symbol next to *100* to bring up the Symbol Selector. Click the Fill Color drop-down arrow, and choose Rhodolite Rose.

Click OK to close the Symbol Selector.

Double-click the symbol next to *200*, and change its fill color to Ultra Blue.

Click OK to close the Symbol Selector.

When you're finished, the Symbology tab should look as follows.

Click OK.

Cerro San Luis is symbolized by the Laguna and Stenner watersheds.

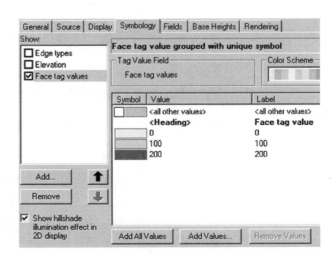

Step 6
Set Vertical Exaggeration

In the Table of Contents, double-click Scene Layers to bring up the Scene Properties dialog. On the General tab, set the vertical exaggeration to 3.

Click OK.

Zoom in and navigate around the mountain. Notice that all the detail of the TIN is retained in the watershed polygons.

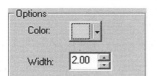

Step 7
Set Base Heights for the Stream Shapefile

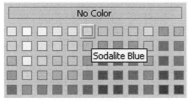

In the Table of Contents, left-click the line symbol under *slo_streams* to bring up the Symbol Selector.

Change the line color to Sodalite Blue, and set the line width to 2.

Click OK in the Symbol Selector.

In the Table of Contents, double-click *slo_streams* to bring up its layer properties. On the Base Heights tab, click *Obtain layer heights from surface* and choose *slo_tin_shed* as the source.

Click OK.

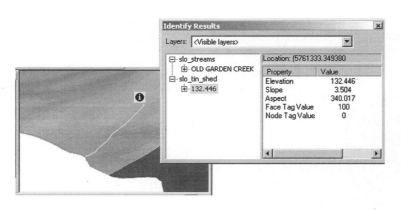

Two things happen: the streams are raised to the elevation of the TIN, but they're also clipped to the extent of the TIN. The only visible stream left is Old Garden Creek, which starts on the northern slope of Cerro San Luis.

There are ways around this, the easiest being to add the streams again as a separate layer, this time not setting their base heights.

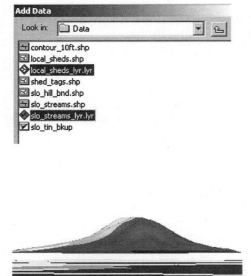

Step 8
Add the Local Stream and Watershed Layer Files

Click the Add Data button, navigate to your *GTK3D\Chapter05\Data* folder, and add *local_sheds_lyr.lyr* and *slo_streams_lyr.lyr*.

Zoom to Full Extent, and then zoom to the extent of *slo_tin_shed*. Take a turn around the data.

Slo_streams_lyr and *local_sheds_lyr* have been symbolized to match the streams and watersheds on Cerro San Luis. Their base heights are set to 0 so that they won't be clipped to any extent other than their own. However, since the TIN starts at an elevation of 110 feet above sea level it floats 110 feet above the layer files.

Step 9
Offset the Stream and Watershed Layer Files

Bring up the layer properties for *slo_streams_lyr*.

On the Base Heights tab, set an offset of 110.

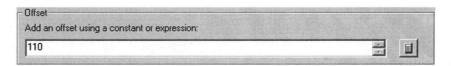

Click OK.

Give *local_sheds_lyr* the same offset.

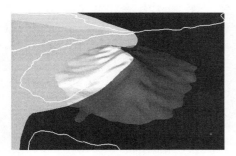

Navigate around the scene. The TIN of Cerro San Luis is symbolized by its watershed boundaries, and the surrounding streams and watersheds are displayed at the correct elevation.

Step 10
Close ArcScene

When you're finished checking out the data, close ArcScene. Give the scene a name of your choice and save it in your *GTK3D\ Chapter05\MyData* folder.

EXERCISE 3: CHANGE TIN SYMBOLOGY AND CLASSIFICATION

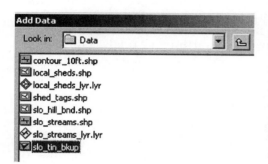

The cells in a raster can hold one type of value only, so when you create a surface raster you have to choose the type of phenomenon it will represent: elevation, slope, or aspect. When you make a TIN, all of that information is included, and you can represent elevation, slope, or aspect just by changing the TIN's symbology. In this exercise you'll change the ways your TIN of Cerro San Luis is symbolized and classified.

Step 1
Start ArcScene and Add Data

Open ArcScene and click the Add Data button.

Navigate to your *GTK3D\Chapter05\Data* folder and add *slo_tin_ bkup* to the scene.

The TIN is displayed by default with an elevation scheme of nine classes at equal intervals, with soft and hard edge types.

Step 2
Change Vertical Exaggeration

In the Table of Contents, double-click Scene Layers. On the General tab, set the vertical exaggeration to 3, and change the background color to a light blue of your choice.

Click OK.

Step 3
Get to Know the TIN's Edge Types

In the Table of Contents, double-click *slo_tin_bkup* to bring up its layer properties. Click the Symbology tab.

In the Show box, uncheck Elevation.

Select *Edge types*, and under the Symbol panel click Add All Values.

Four line types show up (0, 1, 2, and 3), along with their descriptions.

Double-click the line next to 0 (Regular Edge). When the Symbol Selector pops up, choose a thin black line symbol.

Click OK. Do the same for the other three symbols, but choose a fat red line for 2 (Hard Edge) and a thin, bright green line for 3 (Outside Edge).

It doesn't matter which symbol you choose for Soft Edge, because there are none in this TIN. Soft edges are often used to delineate study areas instead of geographical boundaries. The TIN you made in Exercise 2, *slo_tin_shed*, has soft edges defining the watershed polygons, because you triangulated them as a "soft value fill."

Uncheck the box next to *< all other values >*.

Symbol	Value	Label	Count
☐	<all other values>	<all other values>	0
	<Heading>	**Edge type**	**26825**
	0	Regular Edge	24433
	1	Soft Edge	?
	2	Hard Edge	1225
	3	Outside Edge	1168

When you're done, the Symbol panel should look something like that shown at left.

Click OK.

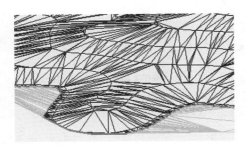

Zoom in and take a look. The hard edge is the line made by the outside contour line, which you used when you clipped the TIN in Exercise 1. The outside edges make up the area beyond the contour line, and the regular edges are the bulk of the TIN. Looking at the edges gives you a better understanding of how the TIN is triangulated.

(Soft and hard edge types are often referred to as soft and hard breaklines in the ArcGIS Help.)

Step 4
Add Slope and Aspect Symbol Schemes

Bring up the TIN's layer properties, and click the Symbology tab.

In the Show box, choose Edge Types.

Click the Remove button.

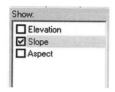

Click the Add button. The dialog that pops up contains a list of ways to symbolize the TIN. You used this in the last exercise to symbolize the watershed TIN by its tag values.

Select *Face slope with graduated color ramp* and click the other Add button.

Also, add *Face aspect with graduated color ramp* and then click Dismiss.

Uncheck the Elevation and Aspect schemes in the Show box, and choose Slope to highlight it.

Click OK.

3D Analyst symbolizes each face of the TIN by its degree of slope. The ranges are classified into nine categories by default, but you can change that. You can also change the color scheme, and individual colors within it.

Step 5
Change Slope Classification

Open *slo_tin_bkup*'s layer properties. On the Symbology tab, highlight Slope in the Show box. (3D Analyst highlights the top entry in the Show box each time you reopen the layer properties. If you like, you can use the arrow buttons to put Slope at the top of the Show box.)

Click the Classify button.

Change the number of classes to 4. In the Break Values panel, change the breaks to 5, 10, 20, and 90.

Click OK.

In the Symbol panel, double-click the symbol for *20 – 90* and change it from red to bright yellow, to distinguish it from the orange symbol above it.

Click OK.

Now Cerro San Luis is symbolized by four slope value ranges. There aren't actually any slope values higher than 20 degrees, but if you had left that symbol red you wouldn't be able to tell it from the orange.

If you like, try to figure out the TIN's highest slope value by using a Natural Breaks classification and increasing the number of classes. Just remember that Slope won't be automatically highlighted in the Show box unless you move it to the top.

Step 6
Symbolize the TIN by Aspect

Open the Symbology tab again. Uncheck Slope and check Aspect.

Click OK.

The default color scheme has eight classes for the eight major compass directions, plus a gray symbol for flat areas. The North symbol is shown twice.

Being almost circular, Cerro San Luis lends itself well to being symbolized by all eight directions, so you won't really improve upon it by changing its default settings. In many cases, however, you might want to simplify the color scheme to show that most areas face one direction or another.

Step 7
Examine the TIN in Orthographic View

From the View menu, choose View Settings.

Change the scene to Orthographic. Move the dialog out of the way so that you can see the scene.

Looking at your data in orthographic view is a good way to be sure of your orientation. Orthographic view mimics the perspective of ArcMap, with North always at the top.

You can see that the ridge line runs roughly east-west. If you change the aspect color scheme, you can emphasize this.

Step 8
Change Aspect Classification

In the Table of Contents, right-click the Northeast symbol.

Also, change the color to Mars Red.

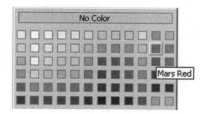

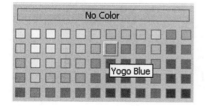

Change the Northwest symbol to Mars Red.

Select the symbol for South, and change it to Yogo Blue.

Do the same for the Southwest and Southeast symbols.

Now it's more obvious that the ridge divides the hilltop into northern and southern directions.

In View Settings, change the scene back to Perspective view, and close the dialog.

Step 9
Experiment with Elevation Classification

Double-click *slo_tin_bkup* in the Table of Contents.

On the Symbology tab, uncheck Aspect in the Show box, and check Elevation.

Change the number of classes, and pick a color scheme that appeals to you. Experiment with the individual colors for each symbol, if you like.

You can create some pretty cool (and even informative) surface features by changing a TIN's elevation, slope, and aspect symbology.

Step 10
Close ArcScene

Feel free to display the TIN with the other types of symbology listed in the Add Renderer dialog. When you're finished experimenting, give the scene a name of your choice and save it in your *GTK3D\Chapter05\MyData* folder.

Chapter 6

3D Features and More Surface Analysis Techniques

Now that you understand the basics of raster and TIN data structures and you're familiar with the workings of ArcScene, it's time to reward yourself with the simple activity of drawing 3D shapefiles in ArcMap. You'll also learn how to calculate 3D area and volume statistics, draw a line of sight along a surface, and create a cross-section profile of that line.

3D Features

We discussed 3D vector features briefly in Chapter 1, so you may remember that 3D points, lines, and polygons are just like their 2D counterparts except for the fact that they store elevation *z* values. A point has one *z* value, whereas lines and polygons store a *z* value at each vertex.

There are two basic ways to create 3D features. You can either digitize them in ArcMap, using a TIN or an elevation raster as the source for *z* values, or you can convert existing 2D features to 3D. When you convert features to 3D, you can get the *z* values from a TIN, from raster surface, or from a field containing elevation values in the feature attribute table. You can also set a single value that assigns a constant elevation to all the vertices in the feature layer.

3D ANALYSIS TECHNIQUES

The white line represents the path water should take from its origin at the top of the slope.

The steepest path tool models the route a liquid such as water or lava will take on a slope. You can also use it to test the accuracy of your surface model. The more accurate the model, the more closely the modeled path of a water source will line up with the course the actual stream bed takes.

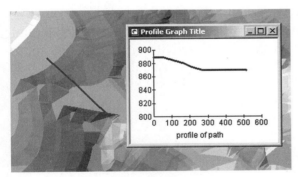

The graph shows the height and distance, in feet, of the line drawn along the surface of the TIN.

Profile graphs show a cross section of a surface along a line you draw, charting the distance of the line on horizontal and vertical axes. This is a function of 3D Analyst, but it can only be done in ArcMap.

A line of sight is related to the targets and observers you studied in Chapter 3, and the viewsheds you studied in Chapter 4. In this case, you choose an observer and a target, and 3D Analyst draws a line directly between them, indicating which parts of the surface along the line are visible to the observer and which are not. Like the profile graph, the Line of Sight tool only works in ArcMap.

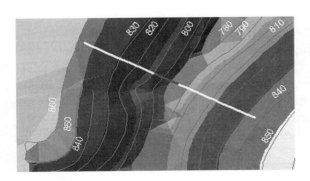

The line drawn across the gully between two points shows not only whether the target is visible but which parts of the surface along the line are hidden from view. In the graphic, the light sections of the line (green) show where the surface is visible. The dark section (red) shows where the view of the surface is obstructed.

Area and volume statistics calculate a surface's 2D area, the surface's actual area (lumps and all), and the volume of material between the top or bottom of the surface and a reference plane you specify. These numbers are useful for applications such as cut and fill, damming, and lake construction.

Now you're ready to take on the last three exercises of the tutorial. You'll convert 2D features to 3D, create a 3D shapefile in ArcCatalog, and digitize in ArcMap. You'll also find area and volume, draw a line of sight, and create a profile graph.

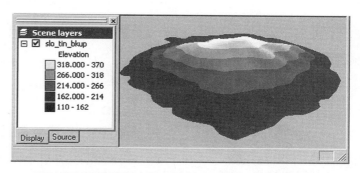

These statistics are calculated for the TIN above from an elevation plane of 200 feet to the top of the hill. Notice that the surface area is greater than the 2D area. Unless a surface is flat, the actual surface area will always be greater than the 2D area.

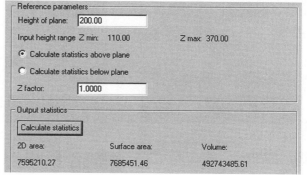

EXERCISE 1: CONVERT 2D FEATURES TO 3D, AND DIGITIZE 3D FEATURES IN ARCMAP

Generally, you make 3D vector features for the sake of convenience—so that you don't have to set their base heights to another layer when you add them to ArcScene. In this exercise you'll convert existing 2D line and polygon features to 3D using a TIN of a rural area near Brushy Creek in Carlisle, KY. You'll also create a new 3D shapefile using ArcCatalog and ArcMap.

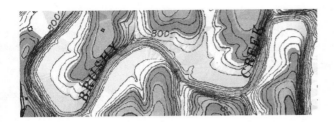

Step 1
Start ArcScene and Add Data

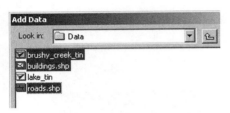

Open ArcScene and click the Add Data button. Navigate to your *GTK3D\Chapter06\Data* folder and add *brushy_creek_tin*, *buildings.shp*, and *roads.shp*.

Step 2
Examine the Data

Navigate around the TIN and the two shapefiles. The roads and buildings are 2D features and don't have their base heights set to the TIN, and so they float below it, as shown at left.

Zoom in and take a look at the depression in the middle of the TIN, outlined here in white. Later on you'll be digitizing a lake polygon to fill this basin.

Step 3
Examine the Shapefile Attribute Tables

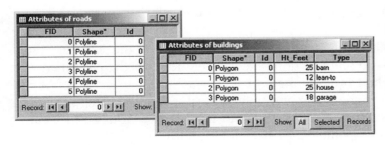

In the Table of Contents, right-click *roads* and choose Open Attribute Table.

Also open the attribute table for the *buildings* layer.

You can tell that the features are 2D because they don't have a z in the Shape field. Notice the *Ht_feet* field in the polygon table. You could use these values to convert the buildings to 3D, but they'd still be mighty short compared to the elevation above sea level of the TIN. So, you'll use the TIN to convert the buildings to 3D, and later you'll use the *Ht_Feet* values to extrude the buildings from the TIN's surface.

Close the attribute tables.

Step 4
Convert the Building and Road Features to 3D

From the 3D Analyst menu, choose Convert, and then Features to 3D.

Examine the dialog for a minute. It looks a lot like the Base Heights tab on a Layer Properties dialog. You're familiar with setting base heights for vector features; you know that depending on what source you have, you can set them by typing in a constant value, by using the Calculator button to assign values from a field in the feature layer's attribute table, by assigning the elevation values from a TIN or raster surface, or by using existing z values in the feature layer's Shape field. Here, you have the same choices—except for the last one, of course, because the purpose of converting features to 3D is to create a new layer that has z values in its Shape field.

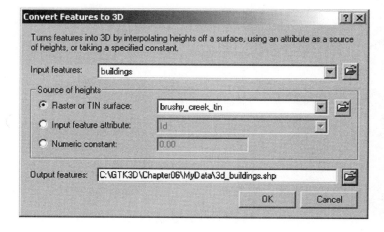

In the dialog box, the Input features should be buildings. Set the source of heights to *Raster or TIN surface*, and make sure it's using *brushy_creek_tin*.

Click the Browse button next to the Output features box. Navigate to your *Chapter06\MyData* folder, and name your new shapefile *3d_buildings.shp*.

Click the Save button. When you're done, the dialog should look like this:

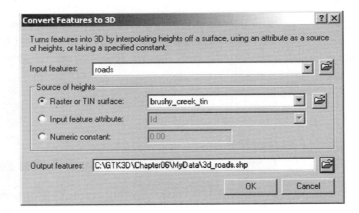

Click OK.

The new shapefile is added to the scene.

Repeat Step 4 for the *roads* shapefile, but name it *3d_roads. shp*.

The new *roads* shapefile is also added to the scene. Now the features lie on top of the TIN.

Step 5
Examine the New Layers

In the Table of Contents, remove the original *roads* and *buildings* shapefiles.

Click the line symbol under *3d_roads*. In the Symbol Selector, change the line width to 2 and choose a light color that will stand out against the TIN.

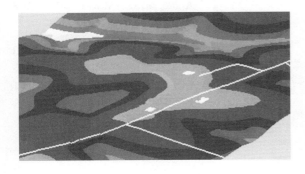

Change the color of the building features, too, if you like.

Open the attribute tables of roads and buildings. Now the Shape fields contain, respectively, PolylineZM and PolygonZM. The Z stands for elevation in this case, although a *z* value can be anything other than *x,y* coordinates. The M stands for measured values, which you get every time you convert 2D features to 3D, whether you specifically ask for them or not.

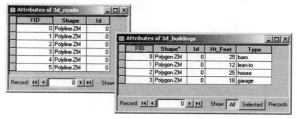

Close the attribute tables.

Step 6
Extrude the 3D Building Polygons

Zoom in to the buildings on the TIN surface.

In the Table of Contents, double-click *3d_buildings* to bring up its layer properties.

On the Extrusion tab, click the Calculator button to bring up the Expression Builder.

In the Fields box, click *Ht_Feet*. It should appear in the Expression box.

Click OK.

Back at the Extrusion tab, make sure you're adding the extrusion to each feature's minimum height.

Click OK.

Now the 3D buildings have the correct base heights, set to their *z* values, and they're extruded to their building height—25 feet for the barn, 18 for the garage, 25 and 12 for the house and the lean-to.

Step 7
Save the Scene

From the File menu, click Save. Navigate to your *GTK3D\ Chapter06\MyData* folder and give the scene a name of your choice. You're going to be working in ArcCatalog and ArcMap, so you can either close ArcScene or leave it open and minimize it, depending on your system resources.

Step 8
Create a 3D Shapefile of a Lake in ArcCatalog

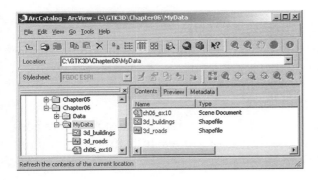

Open ArcCatalog. Navigate to your *GTK3D\Chapter06\MyData* folder.

Right-click the *MyData* folder, choose New, and then Shapefile.

In the Create New Shapefile dialog, change the Name to *3d_lake* and for the Feature Type choose Polygon.

Click the Edit button.

In the Coordinate System dialog, click the Import button.

Navigate to *GTK3D\Chapter06\MyData*. Choose either *3d_roads* or *3d_buildings* and click the Add button. (If you don't have these shapefiles, use one of their backups from your *Chapter06\ Data* folder.)

The Coordinate System panel is updated, showing that you've imported the NAD_83 Kentucky State Plane reference system to your new shapefile.

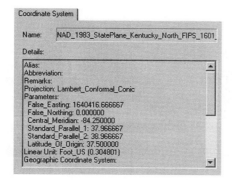

Click OK.

Back in the Create New Shapefile dialog, check the box to give the coordinates *z* values. When you're done, the dialog should look like this:

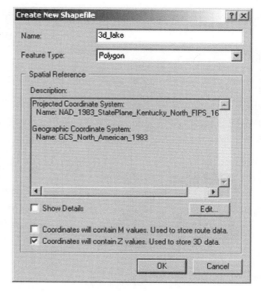

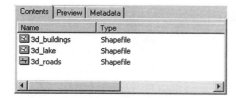

Click OK.

The dialog disappears, and in your *MyData* folder is a new, empty shapefile named *3d_lake*.

Close ArcCatalog.

Step 9
Open ArcMap and Add Toolbars

Open ArcMap.

You may already have the Editor toolbar, the 3D Analyst extension, and the 3D Analyst toolbar loaded in ArcMap, but if not:

From the Tools menu, choose Extensions. Check the box next to 3D Analyst if it's not already checked, and close the dialog.

From the View menu, choose Toolbars. Check the boxes next to Editor and 3D Analyst if they aren't already checked. (Clicking them when they are already checked will uncheck them.)

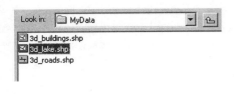

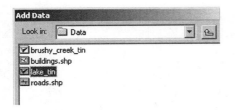

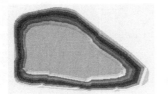

Step 10
Add Data

Click the Add Data button, and navigate to your *GTK3D\Chapter06\MyData* folder.

Add the *3d_lake* shapefile.

The new shapefile is added, but you don't see anything because it contains no features.

Click the Add Data button again. This time navigate up a directory to your *GTK3D\ Chapter06\ Data* folder, and add *lake_tin*.

Zoom to the extent of *lake_tin*.

Lake_tin is a TIN of the lake you looked at on the *brushy_creek* TIN earlier, in ArcScene. You'll use its outer edge to digitize the *3d_lake* polygon.

Step 11
Interpolate a 3D Polygon

From the Editor menu, click Start Editing. The Target in the Editor toolbar should be *3d_lake*.

On the 3D Analyst toolbar, click the Interpolate Polygon button.

Place the cursor in the view.

The Interpolate tools are interesting. Unlike the regular Sketch edit tool, they rely on a 3D surface to get the *z* values for each vertex you digitize. So, when you digitize the polygon it has to lie on or inside the boundary of the TIN. In fact, if during your sketch you wander outside the TIN the Interpolate Polygon tool will automatically snap the vertex to the TIN's boundary. This creates an easy way to quickly make a perfect outline of the lake: if you digitize entirely outside the TIN, no matter how quickly or sloppily you draw, the polygon will snap to the TIN, like so:

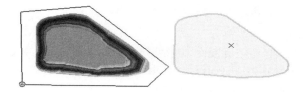

Click once to start drawing. When you've drawn the polygon, double-click the last vertex to finish the sketch. (If you make a mistake, double-click to finish the polygon, and then delete it and start over.)

From the Editor menu, choose Save Edits, and then Stop Editing.

Step 12
Close ArcMap

Give the document a name of your choice and save it in your *GTK3D\Chapter06\MyData* folder. Then close ArcMap.

Step 13
Open the ArcScene Project and Add the 3D Lake

Maximize or reopen the ArcScene project you saved in Step 7. (If you need to reopen it, start ArcScene, click the File menu, and choose your project from the list at the bottom.) It should contain *brushy_creek_tin*, *3d_roads.shp*, and *3d_buildings.shp*.

Click the Add Data button, and from your *GTK3D\Chapter06\ MyData* folder add *3d_lake.shp*.

The lake appears with its base heights set to the perimeter of the basin in the middle of the Brushy Creek TIN. The reason it isn't indented like the TIN you digitized it from is that polygons only carry z values at their vertices, and the vertices are only around the edge. 3D polygons work best for smooth surfaces such as lakes and parking lots.

Step 14
Symbolize the Lake and the TIN

In the Table of Contents, click the symbol under *3d_lake* to bring up the Symbol Selector.

Choose Water Body 3D from the main panel, and then click OK.

Double-click *brushy_creek_tin* in the Table of Contents to bring up its layer properties.

On the Symbology tab, choose a less drastic color ramp, such as Green Light to Dark. (Recall that you can right-click the color ramp and uncheck Graphic View to get the list of color ramp names.)

Click OK.

Step 15
Check Out the New Scene

Now the lake, the buildings, and the roads stand out nicely from the TIN surface.

Step 16
Close ArcScene

Feel free to experiment with symbology, base heights, and extrusion if you like. When you're done, save the document and close ArcScene.

EXERCISE 2: DRAW A LINE OF SIGHT AND A CROSS-SECTION PROFILE GRAPH

In this exercise you'll create lines of sight, draw 3D graphics, and make a profile graph to illustrate the visibility results from the Line of Sight tool.

Step 1
Open ArcMap and Add Data

Open ArcMap, click the Add Data button, navigate to your *GTK3D\Chapter06\Data* folder, and add *brushy_creek_tin* and *lakeview.shp*.

Click the Add Data button again, and this time add *3d_buildings* from your *GTK3D\Chapter06\MyData* folder. (If you don't have this file, use its backup from the *Chapter06\Data* folder.)

Step 2
Symbolize lakeview.shp

Lakeview is a point shapefile that will serve as a visual marker when you draw a line of sight later in the exercise.

Right-click the *lakeview* point symbol in the Table of Contents to bring up the Color palette, and then choose a bright yellow shade.

Step 3
Set Layer Properties for the 3D Buildings

In the Table of Contents, double-click *3d_buildings* to bring up its layer properties.

Click the Labels tab.

In the Label field, choose Type from the drop-down list.

Now click the Symbology tab. In the Show box, click Categories. Unique Values is automatically highlighted.

In the Value Field drop-down list, choose *Ht_Feet*.

Under the main Symbol panel, click Add All Values.

Choose a color scheme you like, preferably something bright that will stand out against the TIN.

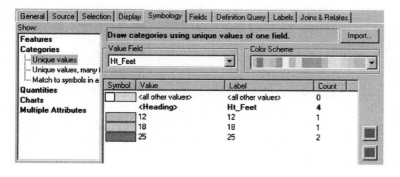

In the Symbol panel, uncheck the box next to < *all other values* >.

When you're done, the Symbology tab should look something like the one at left.

Click OK.

In the Table of Contents, right-click *3d_buildings* and choose Label Features.

Zoom to the extent of the *3d_buildings* layer.

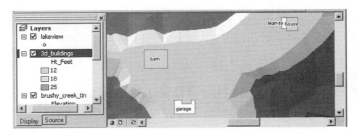

Now the buildings are labeled, and classified by their height in the Table of Contents. This will come in handy when you set the target and observer elevations.

Step 4
Draw a Line of Sight

On the 3D Analyst toolbar, click the Line of Sight button.

The Line of Sight dialog pops up.

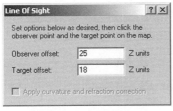

Change the observer's offset to 25. Since the *z* units are feet, this puts your observer at the height of the barn.

Change the target offset to 18 feet, which is the height of the garage.

Now place the cursor on the barn and click once. This sets the observer.

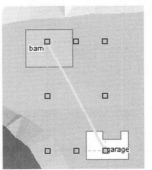

Place the cursor on the garage and click once. This sets the target and draws the line of sight.

Without moving the cursor, look down in the left-hand corner of the status bar. It reports that the target is visible. (If you move the cursor out of the view, the status bar is cleared and you have to recreate the line-of-sight graphic to get another visibility report.) Not surprisingly, the line of sight is green, meaning that you can see the ground along the full length of the line.

Press the Delete key to remove the line-of-sight graphic.

(If you've clicked anywhere else in the scene, or moved the cursor out of the scene, the Delete key won't delete the graphic. In that case, choose the Pointer tool—also called the Select Elements tool—

and click the graphic. When you see little handles around the graphic, you know it's selected and can therefore be deleted.)

Step 5
Draw Another Line of Sight

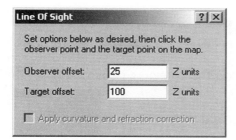

Click the Line of Sight button again, to bring up the dialog.

This time, leave the observer's offset at 25, but change the target's offset to 100.

Click the barn for the observer, and then click the garage for the target.

Again, the entire line is green, and the status bar reports that the target is visible. Although you might think you couldn't see a target 75 feet on a rooftop above you, the Line of Sight tool isn't measuring whether you can see the top of the building or not. Since the target and the observer are on the same plane, 3D Analyst has no indication of any variation in the TIN surface that could obstruct the observer's view of the target. As far as it's concerned, on a flat plane you can see a point in the air 75 feet above you. So, it's always good to do a few tests with the Line of Sight tool, so that you know exactly what is being measured.

Press the Delete key to remove the line-of-sight graphic.

Step 6
Zoom Out to the Lakeview Shapefile, and Use the Identify Tool to Check Points of Elevation

Use the Pan and Zoom Out tools so that you can see the buildings and the lakeview point in the map.

Lakeview represents a bench located on the side of the pond opposite the farm buildings. Before you use the Line of Sight tool again, you should check the elevation of your target and observer.

Click the Identify tool.

Click the TIN surface, right next to the lakeview point. The Identify Results dialog pops up and tells you that the surface there is at an elevation of 850 feet.

Now click the TIN surface right next to the barn. The barn sits at an elevation of 890 feet.

Close the Identify Results dialog.

Step 7
Draw More Lines of Sight

Click the Line of Sight button again. This time, set the observer offset to 4 feet (you're sitting on the lakeview bench) and set the target offset to 25 feet.

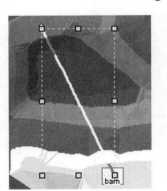

Click the lakeview point, and then click the barn.

The status bar tells you that the target is visible. The line indicates that the slope leading into the pond isn't visible, and that the ground next to the barn isn't visible, either. But the offset of 25 feet makes the barn visible.

Press the Delete key to remove the graphic.

Click the Line of Sight button one more time. Leave the observer offset at 4 feet, but change the target offset to 2 feet.

Click the lakeview point, and then click the barn.

This time, the red and green portions of the line are the same, but the status bar reports that the target isn't visible. The slope leading up to the target cuts off your vision of the bottom 2 feet of the barn.

Step 8
Experiment with the Line of Sight Tool

By changing the length of the line of sight and the target offset, you can get a feel for the cutoff point of visibility. For example, if you change the target offset to 10 feet, and draw the same line of sight from the lakeview point to the barn, the target is reported as visible. (You can see 10 feet up the side of the barn.) But if you extend the line of sight well past the barn with a target offset of 10 feet, the target isn't visible. That's because the angle of the slope leading up to the barn is too steep—a point that will be illustrated more clearly in the next part of the exercise.

When you finish working with the Line of Sight tool, close the Line of Sight dialog. If you have a bunch of graphics left over in the scene, click the Edit menu and choose Select all Elements. Then press the Delete key to remove the graphics from the map.

Step 9
Interpolate a 3D Graphic Line

In the next step you'll create a profile graph of the line you've been drawing between the lakeview point and the barn. Before you do that, however, you have to interpolate a 3D graphic line; the Line of Sight tool only creates 2D graphics.

On the 3D Analyst toolbar, click the Interpolate Line tool.

Click once the on lakeview point, and double-click the barn.

A 3D graphic line is created.

Step 10
Make a Profile Graph

A profile graph is a 2D cross section of a 3D surface along a line you choose. The Profile Graph feature works with 3D lines or 3D graphics, and is available only in ArcMap.

Make sure the 3D graphic is still selected.

On the 3D Analyst toolbar, click the Create Profile Graph button.

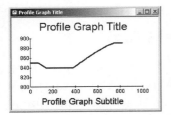

3D Analyst plots the *z* value of each vertex along the selected line to create a profile of the surface at those points. By imagining a direct line of sight from a sitting position on the lakeview bench at 854 feet, you can see that a viewer would be unable to see the slope dropping down to the pond's edge, and also unable to see the ground at the location of the barn, where the surface levels to a flat plane. You can also see that if the barn is tall enough the viewer would be able to see the upper portions of it.

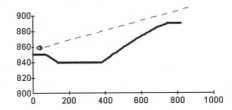

Step 11
Give the Graph a Title

Right-click the title bar of the graph and choose Properties.

In the dialog that appears, change the title to *View over the Pond* and the subtitle to *To the barn, that is.* Or write your own title and subtitle.

Click OK.

Step 12
Change the Name of the Graph

Although the title of the graph was *Profile Graph Title* and you changed it, the name of the graph is also *Profile Graph Title*, and you can't change that fact from the Graph Properties dialog. You have to go into the Graph Manager.

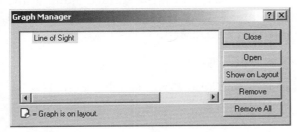

From the Tools menu, choose Graphs, and then Manage.

In the Graph Manager, click once the graph name to highlight it, and then click again to type within the highlighted bar.

Change the name to *Line of Sight* or any other name you prefer.

Click Close.

Step 13
Close ArcMap

Feel free to experiment with other 3D graphics, lines of sight, and profile graphs. When you're finished, give the ArcMap document a name of your choice and save it in your *GTK3D\Chapter06\MyData* folder. Then close ArcMap.

EXERCISE 3: CALCULATE SURFACE AREA AND VOLUME ON A TIN

3D Analyst can calculate the area and volume of a surface above or below an elevation you specify. In this exercise you'll calculate the volume of water in the pond on the farm in Carlisle, KY. The pond is covered with algae, and you want to know how much algaecide you should use to get rid of it without harming the fish, frogs, bugs, and birds that make up the pond's ecosystem.

Step 1
Start ArcScene and Add Data

Open ArcScene and click the Add Data button.

Navigate to your *GTK3D\Chapter06\Data* folder and add *brushy_creek_tin* and *lake_tin* to the scene.

Click the Add Data button again, and from your *Chapter06\MyData* folder add *3d_buildings* and *3d_lake*. (If you don't have these files, use their backup files from the *Chapter06\Data* folder.)

Step 2
Adjust Vertical Exaggeration

In the Table of Contents, right-click Scene Layers and choose Scene Properties.

On the General tab, change the Vertical Exaggeration to 3. This makes the terrain a little more exciting.

Click OK.

Step 3
Symbolize the Lake and the Buildings

In the Table of Contents, right-click the symbol under *3d_lake*. From the Color palette, choose a sickly green shade.

Double-click *3d_buildings* in the Table of Contents to bring up its layer properties.

On the Extrusion tab, check the box to extrude the features. Click the Calculator button.

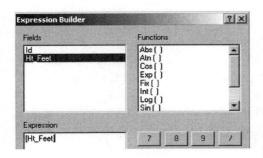

In the Expression Builder, choose *Ht_Feet* from the Fields box. It should appear in the Expression box.

Click OK.

The Extrusion tab should look like this:

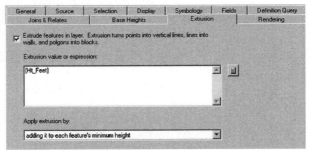

Click OK the Extrusion tab.

Now the lake is algae-colored, and the buildings are represented by their height in feet. Note that they appear taller than they did in Exercise 1. That's because you added a vertical exaggeration to this scene, and it affects extruded vector features as well the surface.

Step 4
Find the Elevation at the Pond's Edge

In the Table of Contents, right-click *3d_lake* and choose Zoom to Layer.

Click the Identify tool.

Click at various points around the edge of the pond, noting the elevation at each location. It rounds to about 850 feet above sea level.

Close the Identify Results box.

Step 5
Examine lake_tin

In the Table of Contents, turn off *3d_lake* and *brushy_creek_tin*.

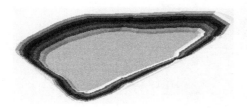

Lake_tin is the surface you used in Exercise 1 to digitize *3d_lake.shp*. Like *brushy_creek_tin*, it was originally created from contour lines. Since the Area and Volume tool calculates statistics for an entire surface, you need to have a TIN or elevation raster of the precise area you're interested in.

Step 6
Calculate Area and Volume Statistics

Since you want to calculate how much water is in the pond, you'll use the elevation around the water's edge (850 feet) as the reference plane, and calculate the volume of *lake_tin* below that plane.

From the 3D Analyst menu, click Surface Analysis, and then Area and Volume.

Change the input surface to *GTK3D\Chapter06\Data\lake_tin*.

Change the height of the plane to 850.

Click the button next to *Calculate statistics below plane*.

Check the box next to *Save/append statistics to text file*. Click the Browse button next to the file path and navigate to your *GTK3D\Chapter06\MyData* folder. Once there, click the Save button.

When you're finished, the Area and Volume Statistics dialog should look like this:

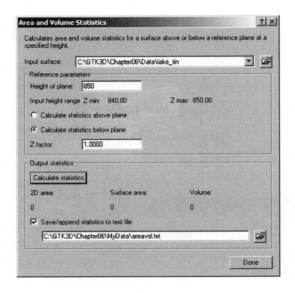

Click the Calculate Statistics button.

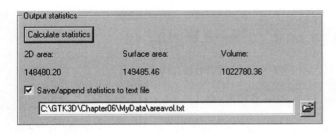

The 2D area, surface area, and volume of the pond are reported in the lower panel.

The 2D area corresponds to the flat plane outlined by the upper edge of *lake_tin*. It's the same area encompassed by *3d_lake.shp*.

The surface area measures the same extent of *lake_tin* as the 2D area, but includes the slopes of the surface. Unless a surface is flat, its surface area will always be greater than its 2D area. A large discrepancy between the measurements indicated that the surface has steep slopes or a lot of variation. To stretch a distance metaphor, you could say that the 2D area is as the crow flies, and the surface area is as the ant crawls.

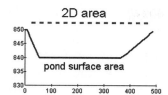

The volume measurement, which is what we're most interested in, is 1,022,780.36 cubic feet.

Click Done.

Step 7
Open the areavol.txt File

Using a text editor of your choice, navigate to the *GTK3D\Chapter06\MyData* folder and open *areavol_txt*. (If you don't have this file in your *MyData* folder, use its backup from your *Chapter06\Data* folder.)

Generally, Notepad works well for these text files, but Word does not. If you're not sure where Notepad is, go to the Start menu on your computer's desktop, choose Programs, and then Accessories. It's usually listed there.

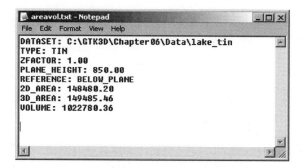

Now you have a printable text file you can use to figure out how many teaspoons of algaecide to use in the pond. (To be on the safe side, of course, you'll use half the recommended amount of algaecide, in case your calculations are off due to an inaccurate surface model.)

Step 8
Change the Color of 3d_lake.shp

In the Table of Contents, turn on all layers.

Right-click the symbol under *3d_lake* and choose a medium blue from the Color palette.

Zoom to the extent of the buildings and the pond.

Now you can envision the farm pond as it will look after you've gotten rid of the algae. (For the record, the ideas in this exercise are for illustration only; the Area and Volume tool should not be used for real applications without being backed up by traditional methods for finding area and volume.)

Step 9
Close ArcScene

Give the ArcScene document a name of your choice and save it in your *GTK3D\Chapter06\MyData* folder. Then close ArcScene, and congratulate yourself. You have finished the entire 3D Analyst tutorial!

INDEX